THE COSMOTHERAPY OF THE ESSENES

THE COSMOTHERAPY OF THE ESSENES

THE COSMIC FOUNDATION OF THE HEALING ARTS

by

EDMOND BORDEAUX SZEKELY

"Gray are all the theories,
but green is the Tree of Life."
—Goethe

MCMLXXXIX
INTERNATIONAL BIOGENIC SOCIETY

SOME BOOKS BY EDMOND BORDEAUX SZEKELY

THE ESSENE WAY—BIOGENIC LIVING
THE ESSENE GOSPEL OF PEACE, BOOK ONE
BOOK TWO, THE UNKNOWN BOOKS OF THE ESSENES
BOOK THREE, LOST SCROLLS OF THE ESSENE BROTHERHOOD
BOOK FOUR, THE TEACHINGS OF THE ELECT
THE DISCOVERY OF THE ESSENE GOSPEL: The Essenes & the Vatican
SEARCH FOR THE AGELESS, in Three Volumes
THE ESSENE BOOK OF CREATION
THE ESSENE JESUS
THE ESSENE BOOK OF ASHA
THE ZEND AVESTA OF ZARATHUSTRA
ARCHEOSOPHY, A NEW SCIENCE
THE ESSENE ORIGINS OF CHRISTIANITY
TEACHINGS OF THE ESSENES FROM ENOCH TO THE DEAD SEA SCROLLS
THE ESSENES, BY JOSEPHUS AND HIS CONTEMPORARIES
THE ESSENE TEACHINGS OF ZARATHUSTRA
THE ESSENE SCIENCE OF LIFE
THE ESSENE CODE OF LIFE
THE ESSENE SCIENCE OF FASTING AND THE ART OF SOBRIETY
ESSENE COMMUNIONS WITH THE INFINITE
THE FIRST ESSENE
THE BIOGENIC REVOLUTION
THE ORIGIN OF LIFE
THE COSMOTHERAPY OF THE ESSENES
THE LIVING BUDDHA
MAN IN THE COSMIC OCEAN
TOWARD THE CONQUEST OF THE INNER COSMOS
FATHER, GIVE US ANOTHER CHANCE
THE ECOLOGICAL HEALTH GARDEN, THE BOOK OF SURVIVAL
THE TENDER TOUCH: BIOGENIC FULFILLMENT
THE DIALECTICAL METHOD OF THINKING
THE EVOLUTION OF HUMAN THOUGHT
THE SOUL OF ANCIENT MEXICO
THE NEW FIRE
ANCIENT AMERICA: PARADISE LOST
DEATH OF THE NEW WORLD
PILGRIM OF THE HIMALAYAS
MESSENGERS FROM ANCIENT CIVILIZATIONS
SEXUAL HARMONY and Raising Happy, Healthy Children
LUDWIG VAN BEETHOVEN, PROMETHEUS OF THE MODERN WORLD
THE FIERY CHARIOTS
CREATIVE WORK: KARMA YOGA
THE ART OF STUDY: THE SORBONNE METHOD
COSMOS, MAN AND SOCIETY
THE BOOK OF LIVING FOODS
SCIENTIFIC VEGETARIANISM
THE CONQUEST OF DEATH

Book Design by Golondrina Graphics

CONTENTS

Dr. Edmond Bordeaux Szekely in 1935

PREFACE
to the 1975 Edition

In 1935, the peaceful mahogany forests on the banks of the Rio Corona, in the state of Tamaulipas, Mexico, had rarely been visited by anyone other than the rural farmers of the region. Lovely and wild, as yet untouched by the ravages of "progress," it was a most unlikely setting for the group of people who arrived there that summer: a large collection of talented young medical doctors, osteopaths, chiropractors and naturopaths, mostly from the east and west coasts of the United States, armed only with tents to live in, and copies of my book, *Cosmos, Man and Society,* 900 pages of revolutionary ideas, already a focal point of controversy in England, where it had been published earlier that year. This was one of the first Essene Seminars, dedicated to a series of my lectures on "Cosmotherapy of the Essenes," and "The Cosmic Foundation of the Healing Arts." It was an unforgettable summer of creative discovery, of intellectual and spiritual awakening, and all who participated were filled with a sense of glad accomplishment at its end: this gifted band of rebels from the healing profession eagerly returned to their work with transcripts of my lectures, later to be published, filled with a new spirit of purpose and dedication. For my part, I was delighted that my transmission of the ancient and ageless Essene message had reached so profoundly so many receptive minds, and the local farmers were happy and relieved that we had not cut down their mahogany forest.

Most of these lectures were published in 1937 by the participants of the Seminar, in a rather voluminous book of 700 pages with the title, *Cosmotherapy, the Medicine of the Future.* Some forty years later, the book is long out-of-print and only a memory. But it seems that a new generation of rebels has emerged, and in this present era when dogmas and rigid ideas in every branch of the healing profession are being questioned, once again a demand for the book exists, and appears to be growing in geometric progression. Fortunately, the most important portions of the book have survived, and one portion has already been published, with the title, *The Essene Science of Life.* The other portion consists of excerpts from the part of the book dealing with the cosmic foundation of the healing arts, or "cosmotherapy," hence the title of this volume. These excerpts are the few fruits remaining from that remarkable summer of 1935. Most of that young band of rebels from a rather tempestuous era of the Essene Movement are gone now, but their influence on this new generation of non-conformist truth-seekers is incalculable.

EDMOND BORDEAUX SZEKELY

San Diego, California, the first of June, 1975.

AUTHOR'S PREFACE

to the 1938 complete edition.

THIS present book is the first complete textbook of Cosmotherapy. I regard it as the synthetic and collective product of the selfless and valuable collaboration of my colleagues. And so, before all else, I wish to thank them for their very valuable contributions, without which this book would never have appeared.

I thank—

The late Mr. Julius Montes, member of our Expedition, who, as secretary of the 1936 Session of our University, translated from Esperanto and condensed this work from my lectures and who sacrificed his life for science at Rio Corona in Mexico.

Mrs. Leone C. Dodson, co-secretary of the above-mentioned Session, who shared with Mr. Montes the whole of this great task and finished it after the death of her colleague.

Mr. L. Purcell Weaver, the translator of those chapters which have already appeared in Cosmos, Man and Society, and the founder and director of the Cosmotherapeutic movement of the British Empire.

Mr. Joseph Fabian, director of the Los Angeles English Publishing Bureau of Cosmotherapy and Mr. George Kosotan, director of the Los Angeles Hungarian Publishing Bureau of Cosmotherapy, who obtained the necessary financial resources for the publication of this book.

Mr. Colman Minarick, the Official Delegate of the International Cosmotherapeutic Expedition in Los Angeles, for his active collaboration in the publication.

Professor Chester Davison of Miami and Mr. Fréderic Marchal of Tahiti for the careful and accurate preparation of the illustrations.

Mrs. L. W. Lusschen, Official Representative of the International Union of Cosmovitalists of Los Angeles, and her friends

PREFACE

Mr. and Mrs. Myron A. Anderson, and Mrs. Axel F. Simonsen, for the heavy and difficult task of proof-reading.

Mr. Sidney White for his devoted and unselfish service.

Mr. I. Port for his valuable collaboration over the material of the book.

Mr. S. Rosenbloom and Dr. Sydney French for their propagandist work for Cosmotherapy in New York.

Dr. Nelson Clanell Oakley, President of Western University, San Diego, for proof-reading and practical assistance in the publication of the book.

Messrs. Colin Campbell and I. Samols for their practical arrangements in connection with the organization of our Summer University at Rio Corona.

Miss Orcella Rexford for her valuable glossary to the work.

Señora Rita de Vargas and Professor Lacalle, the founders of the Mexican Publishing Bureau of Cosmotherapy.

And lastly, I must thank Señor Licenciado Emilio Portes Gil, former President of the Mexican Republic and President of the National Revolutionary Party of Mexico, Mr. N. Draghicesco, Minister of Rumania in Mexico City, Señor Meliton Rodriguez, Mayor of Ciudad Victoria and Mr. and Mrs. Bert Rovere for their intervention and moral support to our Expedition in Mexico against the attacks and intrigues of certain quarters hostile to our work there. Without their help it would not have been possible to bring our Summer University at Rio Corona to a successful conclusion.

This book has been compiled in the leisure hours of those who have laboured together to produce it—after their daily work in business or profession, so the kind indulgence of the reader is asked for any mistakes which may appear. These will be carefully corrected in the second edition.

And now I leave this book to its course, in the hope that it may guide as many as possible into the path of health, regeneration and length of life, and that it may serve as a sure and complete guide to all who seek the truth in the midst of the great chaos existing among the various therapeutic systems and world concepts.

Dr. Edmond Székely.

London, 5th., January, 1938.

The following Introduction by Dion Byngham was written in 1936 as a review of the original *Cosmos, Man and Society*, by Edmond Bordeaux Szekely, with original drawings by Arthur Wragg (sample above). The current publication of the International Biogenic Society with the same title is not at all the same text as the original *Cosmos, Man and Society*, described in the following pages, but is instead the text of another early work of the author, *Medicine Tomorrow.* The unwitting cause of this confusion was Dr. Szekely himself, who liked the title so much that he gave it to two of his books.

INTRODUCTION

COSMOTHERAPY AND PANEUBIOTICS THE QUINTESSENCE OF NATURAL THERAPY AND AN ENCYCLOPEDIA OF LIFE

(Appeared in the greatest Health Magazine in the European Continent: "Health And Life"—London, July 1936.)

By DION BYNGHAM

If the contents of a book could save our contemporary world, which is eddying and crashing to ruin, that book, I believe, has been written. If the contents of a book could redeem us as individuals from a common destiny of disease and death, that same book has appeared. If between the covers of a book might be found the sure path to creative peace, to superb health and beauty, to an optimal abundance of *Joy in Life,* let us open and read. For if, while reading, we should perchance resolve to try and *live* what we read, all these seeming miracles might assuredly happen.

How and where shall one begin? As a *reader,* certainly, at the *beginning,* meditating and mastering every word to the end. But as a reviewer, the significance of some 800-odd pages to compress within four or five ... ! For nothing less than *Cosmos, Man and Society* is the theme.* And the author, Professor Edmond Székely, is a philosopher, a sage and a scientist of immense erudition who—albeit with perfect lucidity and simplicity—has well-nigh taken the reins of infinity within his grasp. Primarily a research-ethnologist, he has been directed in all his

*Cosmos, Man and Society, by Edmond Székely, translated by L. Purcell Weaver, drawings by Arthur Wragg. The C. W. Daniel Co., London. "University Course of Cosmotherapy and Cosmovitality" by Edmond Székely, translated by L. Purcell Weaver and Julius Montes. With hundred illustrations. Published by "International Cosmothepeutic Expedition", Los Angeles, California (U. S. A.)

researches by the light of one dominant motive: to discover and formulate for himself and his fellowman the laws of an optimal life, to focus the entire forces of Eternal Life for the ultimate conquest of death.

As a pre-eminently realistic and practical outcome, Professor Székely has arrived at a comprehensive re-statement of Natural Therapy which excels in completeness, applicability and precision anything so far achieved. And so, as promised in his preface, his book becomes virtually "the encyclopaedia of a new science—Paneubiotics or Omnilateral Aristology—whose purpose is to realize the best and optimal form of omnilateral life based on the totality of the natural and cosmic laws and their manifestations in the individual and society (cosmic, solar, terrestrial and human radiations)."

COSMIC RADIATIONS

In the word "radiations" is really the core and gist of the matter. We live as witnesses to the dawn of the radioactive era. It is therefore essentially in tune with the times that natural therapy itself should be synthesized around the conception of *radiations,* cosmic, solar, terrestrial and human. This, in the first place, is what Professor Székely has accomplished—not, however, as a merely metaphysical theorist discoursing vaguely on "vibrations," but with all the authority of a realistic research-scientist whose findings have been tested and confirmed in every detail by his therapeutic practice and experience in many parts of the world.

It will unquestionably be of supreme interest to the practical reader, whether lay or professional, to find in this book the whole rationale of sun, air, water and earth therapies and hygiene, as well as the properties of properly grown living foods like uncooked fruit and vegetables, re-interpreted and applied in a new light as accumulators and transmitters of cosmic radiations for the healing, health and optimal vitality of man. The entire technique of hydrotherapy (baths, packs, compresses, the internal douche and so forth) is incidentally restated and brought up to date. This makes a valuable section for all who wish for a working knowledge of water treatment, which, in many respects, was the original basis of nature cure.

Dietetic science is likewise fully re-formulated with special reference to the organic salts, vitamins and aromatics as basic constituents and stimulants of the life-processes, as transmitters of vital vibrations to glands and organs and to all the living cells of the body and brain. Fifty foods, "accumulators of cosmic energies," are detailed, together with specimen daily menus for each season of the year. Of several particulars that impressed me about these menus I must mention their extreme simplicity and frugality—yet it is obviously the frugality of a man who *enjoys every* mouthful he masticates with unperverted taste and unspoiled piquancy of appetite. And the stress laid on aromatics: onions, garlic, chervil, chives, shallot, mint, parsley, thyme, celery and the rest.

These aromatics are undoubtedly nature's own *apéritifs.* Through both taste and smell and by virtue of their volatile essences they stimulate digestion, in addition to their valuable, antiseptic and blood-cleansing properties. Professor Székely, who everywhere insists upon the importance of keen healthy *senses,* practically identifies such imponderable neural excitations from our food—transmitted through sight or through taste and smell—with the vitalising vibrations which *are the vitamins themselves.* Many a person, probably, remains obsessed by the blandishments of cunning cookery and dependent upon the *condimental* savour of flesh foods, solely through ignorance of these superlatively natural flavourings which would minister to his instinctive craving for appetising aromas and palatable meals. The raw diet regimen would never be dull were this understood, and no one can be a true initiate of Professor Székely's "natural optimal dietetics" by whom it is not understood.

To trophotherapy and dietetics is added a section describing fifty medicinal plants and herbs with their properties and remedial uses, so that the gist of botanical therapy is also included. The regenerative role of fasting fills a fundamental place, as also does deep rhythmic breathing, since Professor Székely regards sun-irradiated air as the medium for direct transmission of solar and cosmic vital forces through the lungs to the body and mind. From this point of view the yoga doctrine of pranayama receives corroborative evidence. A practical chapter on the Art of Breathing, included in the section on "Individual Harmony," is a revelation in itself. Air, as well as water, must, for instance, have been recently sun-irradiated to be of any living value at all. "Man's roots are his lungs."

CELL-REGENERATION IN SEVEN MONTHS

All of these practical measures and many others are combined in a complete system of what Professor Székely calls "omnilateral cosmotherapy." This term "omnilateral," connoting as it does extension along, and convergence from, *all sides* and *all fronts* as correlated to the central objective of optimal healing, health and life, is what distinguishes Professor Székely's truly "cosmic" conception from all limited, "unilateral" or one-sided methods hitherto applied. As a therapeutic system, competently administered and with the faithful collaboration of the patient, he has proved that it will not only cure specific and chronic diseases but will actually regenerate the individual entirely, through intensive cell-renewal, in from three to seven months. Adopted thereafter as a way of life, anthropocentric "cosmovitalism" becomes the basis of Paneubiotics, the art and science of optimal vitality and integral wellbeing which, Professor Székely believes, will lead to maximal longevity and even to the ultimate conquest of physical death.

Having said so much I am barely at the beginning of any adequate disclosure of the wealth that this book contains. In fact I have *left out* the beginning, plunged into the middle, and thereby postponed all reference to what for me personally, along with the end, form the most fascinating features. To those who, like myself, arrived at "naturism" and natural therapy not primarily through concern with illness but as the practical outcome of a sifting of cultural values and intuitive philosophy of life, by far the most interesting aspects will relate to how Professor Székely reaches his conclusions and the logical consequences to which he carries them.

ATOMS AND STARS

To the whole vast sidereal, astrophysical and astrochemical purview, the correlative etheric and ultra-etheric, ultra-atomic and intra-atomic vibratory *milieux* through which Professor Székely traces life in evolution from nebulae to man, I can make but a passing and meaningless reference here. That man is one with the cosmos, a microcosm with a macrocosm, has been taught by the seers through immemorial ages and notably by Edward Carpenter and D. H. Lawrence among poet-philosophers of recent times. It has remained for Professor Székely to give scientific formulation to that "eternal ocean of cosmic radiations" whereby solar systems and atoms are reciprocally related and the cells of man's body and the thoughts of his brain vibrate in tune with the furthest stars.

"The most elementary dynamic formations are those of matter (matter being merely an aggregate of the movement of atoms), *while the highest dynamic formations are those of currents of thought.*" So the long history of this vertiginous dynamism of thought itself is reviewed by Professor Székely in terse summaries of nearly a hundred philosophers from Thales down to date. For as part of his omnilateral system is included "the entirely new and original application for therapeutic purposes of the masterpieces of universal literature, philosophy and the arts based on psychotechnical diagnosis of individual natures." To Professor Szekely's own philosophic formula of "dialectical correlativism," in which is woven really the whole fabric of this thesis, I cannot do justice in this outline.

MAN AND TREES

There remains still to be noted the ethnological and ethnographical backbone of the book. This—more than the others—has been Profesor Székely's especial line of research, in which he has made some original and significant discoveries. As a pragmatic consequence he is able to correlate all that is best in the most recent scientific and cultural developments with a modernistic renaissance of the heliolithic culture of *homo sapiens sylvanus,* the original fructivorous man of the forests and trees. It was from this dominant species, characterised by maximal longevity and endowed with a cerebral capacity for unlimited intelligence, that subsequent culture and civilization took its source. Identifiable with the Cro Magnon family, with the prediluvian patriarchs of the Old Testament, and with other legendary figures of the earliest folklore, this great race received its first severe set-back from that planetary and climatic disaster recorded as "the Flood." This destroyed its first optimally favourable natural *milieu* and caused its scattered survivors to become more pyrolithic (fire-using and food-cooking) and omnivorous (flesh-eating) in their habits—and much shorter-lived. It might almost be said that, dating from that event, a progressive series of enforced adaptations, complicated by human follies, has brought us down to "the present pessimal *milieu* of *homo sapiens faber"—the* office and factory age of treeless cities and towns we exist in to-day.

Trees and forests, declares Professor Székely, "are accumulators of the cosmic forces and radiations and are *the generative source of cerebral energies."* (Italics mine.) Reflecting that it was the forest sages of India who wrote the profound and unsurpassable Vedas and Upanishads, one can well believe that this is so. Forests, moreover, afford the best protection against storms, floods, droughts, extremes of heat and cold, by acting as natural regulators and distributors of the moisture and temperature of the air. Modern man, in his commercial, money-maniac blindness, has devastated and laid waste the green earth. It is a ruin compared with what it used to be, and man himself—"is a veritable bundle of rags and bones compared with the perfect, statue-like forms of his paleolithic sylvan ancestors". The more human beings have destroyed the forests and cut down the giant trees, the more frequent and terrible have been the calamities, storms, hurricanes and cyclones that have descended upon them, raging in recent years with unparalleled violence. It is cause and effect. Yet still must walnut trees be felled to make rifle butts, and miles of forests be sacrificed to provide paper for the *Daily Mail!*

Those of us who have always loved trees will be wholeheartedly with Professor Székely in his campaign to combat every plan for destroying and cutting them down, to promote re-forestation over the whole earth and restore "the Holy Alliance between Men and Trees." Our "mere sentiments" on the matter are, after all, radically justified, just as they are when along with Professor Székely we deplore the "senseless stupid men who relentlessly hunt down little birds, slaying them without pity and without remorse"—with the result that crops are ravaged by vermin and men themselves are plagued by mosquitos and flies. (News now comes to hand of a modified death-ray soon to be commercialized for the destruction of vermin and insects and, doubtless, of *more* little birds. The pious hope is expressed that it may never be amplified and turned on the human insect himself!)

WORLD CHAOS AND CALAMITY

Yet it would be futile to denounce the world's evils—alas! as so many *still do*—with no clear recognition of their root causes in the world-wide economic crisis and crashing chaos of which they are all merely symptoms. It would be equally pointless to evolve grandiose and utopian conceptions of "omnilateral aristology," "paneubiotics" and optimal wellbeing without facing the fact of a corrupt financial system that can only spell their frustration. Professor Székely is undoubtedly a man of great vision, but he is much too penetrating and uncompromising a realist—and much too good a natural therapist—to be guilty of either defect. His is no idealistic flight into phantasy. He sees through to *causes*. His analysis of the present world situation is among the most masterly I have read. Multiplying unemployment, poverty amidst plenty, the unprecedented crisis of "over-production," the impending fearful collapse of finance, the false foundations and futility of the League of Nations- the pell-mell preparations for war (which is only held off for the moment by the very complexity of imperialist antagonisms), slave-mentality everywhere, confusion, jobbery, and antagonism among the ruling classes (themselves but slaves who lord it over slaves), the desperate attempts through dictatorships to bolster the *bourgeoisie,* the whole piled up impossible impasse

—all this is but "the prelude to the frightful disaster which is soon going to overwhelm humanity in all its intensity." Only, it would seem, after terrible catastrophe, can come the "Great Renaissance," for which, however, in the name of humanity we must even now prepare.

THE GREATEST RENAISSANCE

We must, in the words of Romain Rolland, "unite all the spiritual forces of Life against the forces of Death." In that spirit, too, we must utter "an unqualified 'No' to war." We must vow our devotion to "the Optimal and Omnilateral Restoration of All Things." By extending into Society and social behaviour the paneubiotic law of the healthy body, whereby every organ functions for the wellbeing of every other organ as bound up with its own, we must combine to build the kingdom of active love upon earth. Professor Székely's remedy is, again, no regress into "back to nature" primitivism. For him "naturism" and individualism are necessarily correlated with all that is truly valuable in community and technological achievement. He believes in the future, yet his motto is "always everywhere, Here and Now."

And so when, in conclusion, Professor Székely outlines his plan for the new order of self-supporting, co-operative, paneubiotic garden-communes it is an eminently realistic and practical plan, and, as far as I can see, about *the only way out.* None the less, on that account, will these communes be devoted to realization of optimal individual and social wellbeing through the practice of active love, the omnilateral pursuit of knowledge, unison and harmony with the eternal currents of cosmic life. Professor Székely has followed his cosmotherapeutic premises to their logical conclusion, not even faltering before the menace of a world-disease, imbecility and impending disaster which may soon reduce all personal aches and pains to pitiful insignificance. Because of that alone his book would tower like a giant above all the smug little panderings to hypochondria which are too often described as "health" literature!

This book, as he says at its beginning, will surely "start a great selection of individuals"—the fittest to survive in the new paneubiocracy that is to be. Which of us in our present state dare fancy himself among them? Only by their fruits shall they be known!

The colossal task of translation has been nobly accomplished by Mr. L. Purcell Weaver, but for whom, declares the author, this book would still be lying in manuscript form in French among the records of his various scientific expeditions.

CLASSIFICATION AND CHARACTERISTICS OF THE COSMIC RADIATIONS

(Summary Table)

Wavelength (in cms.)	*Nature of the radiations*	*Their influence on the atoms*	*Temperature (in absolute degrees)*	*Where found*
$7{,}500 \times 10^{-8} \gg 3{,}750 \times 10^{-8}$	Visible rays	Bombard external electrons	$3{,}850° \gg 7{,}700°$	In the stellar atmosphere
$250 \times 10^{-8} \gg 10^{-8}$	X-rays	Bombard internal electrons	$115{,}000° \gg 29{,}000{,}000°$	In the interior of the stars
$5 \times 10^{-9} \gg 10^{-9}$	Soft γ-rays	Discharge all or almost all the electrons	$58{,}000{,}000° \gg 290{,}000{,}000°$	In the central portions of the dense stars
4×10^{-10}	γ-rays of Radium B	Bombard agglomerations of nuclei	720,000,000°	In cosmic space
5×10^{-11}	The shortest rays	Ditto	5,800,000,000°	Ditto
13×10^{-13}	The most penetrating rays	Destruction and creation of concomitant protons and electrons	2,200,000,000,000°	Ditto
?	Vital rays	Attraction of radiations in affinity	?	In living organisms
?	Mental rays	Ditto	?	In cerebral systems
?	**Unknown rays**	?	?	?

INTRODUCTION TO THE INTERNAL COSMOTHERAPY THE COSMOVITAL RADIATION IN THE CELLS, GLANDS, HORMONES AND VITAMINS

Waves, Radiations, Vitamins

Vitalizing Reconstituents

The human body is daily disintegrating in all its constituent parts, to be entirely renewed every seven years; so that a man who reaches the age of eighty, for example, has been reconstituted nearly a dozen times. In order, then, to effect this complete septennial reconstruction, to which must be added the restoration of losses due to molecular expenditure occasioned by work, the amount of materials disintegrated must be replaced each day by an equal amount of other new materials. Since the human body requires not inert materials, like a machine, but materials adapted to its composition, that is, living bodies suited to this role of restoration, it is impossible logically to conceive of the supply of food as consisting of substances other than those possessing the same vitality as our bodies. Material life is really only a perpetual process of cellular interchanges between the cells obtained in the foods we take and those of the human body that time and wear have destroyed, the cells of the foods being transmuted into those of the body.

Regeneration of the Cells of the Body

It must not be forgotten that everything in our bodies is composed of cells, and that these are transmitted by the blood which possesses two types of them, red corpuscles and white corpuscles, or leucocytes.

The blood extracts the cells from the food and distributes them as required for the restoration of the flesh, bones, cartilage, epidermis, marrow of the various nerve substances and the nerves themselves, including the hair on the head and body, and the nails, which are all dependent on the cellular structure. These myriads of different cells necessarily require for their perpetuation in the human body contact with the elements of regeneration contained in foods. As these protoplasmic cells are living beings in a state of evolution and reproduction like everything alive, whether vegetable or animal, it follows that extreme degrees of heat and cold destroy them just as they destroy the two latter forms of existence. The result is that if we eat nothing but cooked foods, we lose the majority of these cells, all of which cannot resist the average temperature of cooking (100° Centigrade), so much so that, through failure in the supply of new cells, the body will develop weakness in the organs where there is a deficiency. This leads to a destruction of the normal balance of our structure and will eventually become one of the causes of disease. If this process is extended and we feed on nothing but preserved foods sterilized at a temperature of 120° Centigrade, we meet with disaster, since every part of the body will become decrepit and finally die when the cells have been destroyed through failure to renew themselves.

Living Cells Require Food

In connection with the problem of our existence, we must also consider, besides the action of stimulation and vibration performed by living foods, the part played by the cells in the building up and regeneration of the complete structure of the human body. We see that right from birth, our whole body: skin, hair, flesh, fat, bones, is formed only of cells, and not of chemical bodies capable of analysis. It is formed of living cells, each of which has a life of its own and a part adapted to the life of the organ that each kind of cell is entrusted with maintaining in good physiological condition. By furnishing new cells, it also has to regenerate the organ in question and so prevent its premature decrepitude.

This cell, which is the foundation of our organic structure, being alive, needs foods in order to live. It is not just any food, or always the same food that it requires; it must have foods which are in harmony with its own composition and with its particular function in the continual restoration of every part of the human body. The foods must be varied and composed of cells as closely allied as possible to the composition of the organs that these foods are designed to regenerate. It is necessary that the foods, however well varied they may be, from which we propose to draw the new replacing cells, should be alive. Otherwise, it is impossible for the regeneration of the cells to occur, since there are no longer any living cells in dead foods to bring about the regeneration of our living organism. The inevitable consequence is that decrepitude of the human structure ensues, beginning with the organs whose cells of replacement are missing in the foods.

Microscopic examination of one of the cells which form the basis of every organized structure, whether of the animal or vegetable kingdom, leaves one astounded by the ingenuity which is shown in their constitution. There exists in the cell, in embryonic state, all that is necessary both for its structure and for the vitality of that structure: for instance, in the case of the typical cell, we find a nucleus with a coiled-up thread inside, the entirety of which forms the vital center entrusted with carrying out, in particular, the phenomenon of nutrition. Around the nucleus are vacuoles containing albuminous juices, cellular protoplasm and nuclear juices. The whole of this outside portion is entrusted more particularly with the functions of sensibility, as it is especially noticeable in the leucocytes and nerve cells. In each cell, then, of any organ, there will be vacuoles containing protoplasmic juices, which will spread through our organism and will become stationed in the places for which each of these juices is intended. In these substances enclosed in the cells, in a state of colloidal division, we find every material necessary and admirably organized for the building up of the entire human structure. Each cell provides the organ which it constructs with the kinds of materials particularly required in the way both of minerals and of nuclear juices (the name given to the hundred or so forms in which albumen can appear). These albuminous juices participate in the construction of the human building as a binding substance or mortar designed to make a well-amalgamated cellular mass, which constitutes either a muscle, a bone, the epithelial surface of a fascicle, or a hair, etc. In every part of our body there is albumen, which in chemistry is termed nitrogen. Nature provides us with a wonderful model of cells on a large scale, containing all the elements necessary for the construction of an animate being. Such a tangible and visible cell, with all the substances accumulated for the creation of life and the structure of another life, including the coiled-up nucleus, is to be found in the egg; in the germ the complete likeness is accumulated on a small scale, with its hereditary instincts derived from its ancestors. The cell will be classed among constructive protoplasmic cells, defensive leucocyte cells, or animating nerve cells, according to its function and purpose.

A cell is an element in which, besides carbon, oxygen, and hydrogen, nitrogen predominates, while, in the red blood corpuscles, iron and manganese are found as well. In the leucocytes (white corpuscles), in addition to nitrogen, we find traces of magnesium and calcium; in the neurons (nerve cells) traces of phosphorus, and that is all: nothing else besides the central vacuole and its coiled-up thread (center of the vital action of the cell) and the cellular protoplasm and the nuclear juice vacuoles, source of our regeneration. In fact, this marvelous little mechanism, which is the source of morphological life, is in final analysis nothing but a few dead ashes. Man, the masterpiece of nature, is exactly the same after cremation. In nature there are generally only organized bodies, each endowed with a chemical and physical life. All their elements live in symbiosis side by side in mutual harmony as long as no abnormal occurrence upsets the marvelous balance of every natural body, whatever it may be, chemical or living.

Such disorganizing influences are fire (which nothing can resist), and numerous chemical products, including alcohol. Cells with the greatest resistance are destroyed between 100° and 115° Centigrade, while most are destroyed between 60° and 70° Centigrade in some cases, or even 45° to 50° Centigrade in others. We mention this to stress the difference between living food composed of a minimum aggregate of regenerative cells and dead food totally unprovided with them.

ORGANIC MINERAL SALTS AND THEIR VIBRATORY ACTION UPON THE LIFE OF THE CELLS

Everything in our bodies is so admirably organized for life in symbiosis (vital harmony), that if one of its components, cell or chemical or organic element, is missing or deficient in quantity, there is immediately a correlative disequilibrium of our vital functions in the shape of organic disorders. This disequilibrium will be only partial if the component is only partly missing, but if completely missing, the disequilibrium will be total.

The cells of which all our organs are composed, require for life and reproduction different supplies of organic mineral salts for each kind of cell. The presence of their vibration (the basic characteristic of the metalloids) is essential to the existence of each of the cells. They need not only one mineral but the sum total of the minerals and metalloids of the earth in varying proportions, with a preponderance of phosphorus, magnesium, lime and iron. The nerve cell, in order to live, reproduce and regenerate itself, needs phosphate of magnesia as its principal constituent mineral salt (in the form of organic salts and not of chemical phosphorus and magnesia), such as is found in certain foods and particularly in whole grain cereals and breads. However, if this salt is totally or partially lacking in food, the nerve cells will wither and the whole or part of the organism will fall into decay. There is one and only one remedy for this: an internal supply, in the form of food, of phosphate of magnesia, because salts of phosphomagnesia are the only ones whose irradiating characteristic can vibrate in harmony with the necessities and particularities of the life of the nerve cells. Take another example: when the red blood corpuscles lose their vitality, the general condition at once grows poor and the whole body is impaired. This is the classical anaemia. If the red blood corpuscle is provided with what is missing, and regains its full power by being supplied with iron salts vibrating in harmony with the corpuscles without which they cannot live, the vital equilibrium of these cells is at once re-established. These salts are found in almost all vegetable foods (generally in those which are green in color) and in very appreciable amounts.

As a further example, we may mention the principal mineral salt derived from lime, whose vibratory interference is necessary to maintain the activity of the cell entrusted with the defense of the human body, the leucocyte cell or white corpuscle. We can see from this how important it is that every constituent cell of the human body should contain the organic mineral salts which can be supplied only through regular absorption of food. When we speak of organic mineral salts, we mean living, and consequently vibrating, salts; that is to say, those that are only to be found closely blended with natural foods: in the structure of plants and of fruits, and also in that of eggs and of milk.

The three examples given, which apply to the three essential cells, could be multiplied for every organ of the body merely by changing the name of the organic mineral salt, since each organ always has a cell as the foundation of its constitution. But we should also realize that the cells need the molecules of other mineral salts besides their special one. For instance, the cells which have a salt of lime (calcium) as one of their bases also require salts of silicon (silicium) to secure sufficient elasticity and salts of fluor (fluorine) to provide the necessary hardness to the cells especially charged with making up the external parts of the bones and the teeth (periosteum).

The same applies in our epidermic cellular system (skin and the surfaces of all our internal organs) and in the capillary and ungual systems (hair of the head, hair of the body, nails), etc. Each of the various kinds of cells at the base of its structure needs the presence of a mineral salt in order to live, or of several mineral salts whose vibratory particularities are able, by harmonizing with the vibration proper to each of the cells, to maintain a perfect state of equilibrium in all these cells which live in us, as in a wonderfully arranged and organized zoological garden, each in its place, with a well-defined function, giving one another mutual aid and support. As long as this accord is maintained, the human structure continues to enjoy stability and health.

If our cellular system loses its harmony through a predominance or scarcity of one of the kinds of cells, we immediately experience a disequilibrium of our vitality in the shape of disease of the organs affected.

Confining ourselves to the three kinds of cells (protoplasmic, defensive and nervous), on which the constitution and morphological equilibrium of our organism rests, we can make the following statement: If one of these three cells proceeds to disappear, then the morphological product generated by the two remaining cells in the absence of the third, can only be a monstrosity, a kind of parasitic growth without plan or purpose, existing at the expense of everything around it. In short, the two cells, disharmonized as a consequence of the disappearance of the third, are the source of the steadily growing scourge known as malignant tumor or cancer, which has its origin in the unbalanced condition of the two cells disharmonized by the disappearance of the third.

A little reflection will show us that the absence of organic mineral salts (of the salts endowed with a vital vibration without which no cell can live) is due to the foods which contain them having been sterilized by overcooking. The extreme heat breaks up and dissociates their living organic mineral salts, and they become dead mineral salts. In other words, they have become chemical salts incapable of supplying the cells with vital stimulation, the vibration without which their life, reproduction and regeneration are impossible.

We may add as the second principal cause of food deterioration, the presence of alcohol. Alcohol is one of the greatest enemies of life. If taken habitually, even in small quantity, it reacts on the life of the organic mineral salts and also on that of the cells through contact with them.

CELLULAR RADIATION

The influence of the radiations of the cells, which are really vibrations or irradiations, makes itself felt upon the cells of the surrounding tissues in the course of formation.

If a group of cells has become incapable of radiating, and as a result, incapable of exercising its vibratory influence with respect to another neighboring group which is growing up, the latter group, through failure to receive this vibration which is indispensable for its growth, remains incapable of living and consequently has to disappear. The irradiating influence from cell to cell is the most important discovery so far made in biophysiology. The existence of these vibratory influences in connection with the life of the cell, and thus of every living thing, gives the most tremendous knockout blow to the theory of analyzable matter that it has ever received.

Everything which exists on the earth is simply the product of vibrations and radiations. These vibrations which we have classified in the spectrum of solar light where they originate, are so well linked and harmonized with each other that the absence, or merely the attenuation of one of them *ipso facto,* generates disturbances, the effects of which emerge as a disequilibrium in function among our various organs. The latter are naturally attuned to each other, with the result that decay in one immediately generates hypertrophy in another.

It follows from what we have just said that *the secret of health and longevity is to be found in harmony between the cellular vibrations of the organism and the cosmic radiations.*

GRAFTING OF ENDOCRINE GLANDS, BLOOD TRANSFUSION AND TRANSMUTATION OF CELLS

Since human organs, as well as the cells, have a different constitution in each individual case, we must not rely on the organs or cells of other people, and *a fortiori* on the organs or cells of animals; we must not depend on rejuvenating ourselves through their transposition by means of grafting or transmutation by unnatural, external means. If such a method were within the realm of possibility, we should have to admit that nature has made a mistake in one of her works. As it is impossible for nature to be in error, it is man who once again is mistaken, through the sin of pride.

We have forgotten that complete old age is merely the sum total of a mass of partial old age and that the most powerful of hormones cannot rejuvenate the entirety of the decayed cells and senile organs. We also forget our *farouche* individuality and the veritable xenophobia of our cells, which quickly eliminate the foreign grafting, not only when it is derived from a neighboring species but even when it is taken from a subject of the same species. So if you wish to become young again when you feel yourself falling prematurely into decrepitude, do not rely on rejuvenation baths, or on graftings of young organs previously removed from monkeys, or again on blood transfusions containing young cells, practices as unnatural as they are immoral. But rely with every confidence, without risk of error or harm to your general condition, on the means so generously put at your disposal by nature; you will never have to regret using them because it is not a question of partial reform but of doing entirely the reverse of what you were doing.

For a successful rejuvenation cure, all we have to do is exclude from our diet everything that is pernicious and harmful to the health of our organs and replace it by a suitable selection of the appropriate living foods; that is, foods rich in life-generating, or *biogenic* materials, which constitute the cells of replacement; provided always that these cells of replacement are still living and are not the corpses of cells, as is the case when foods have been sterilized in an autoclave or mortified by contact with alcohol. These dead cells, instead of nourishing our organs, clutter up our bodies with toxic substances which, in the long run, destroy us.

HORMONES, VITAMINS, WAVES, RADIATIONS

The Glands of Internal Secretion

The glands of internal secretion (endocrine glands) play a part in our existence, both separately and in the coordination of their functions, which is of capital importance for the maintenance of our health in proper balance. The principal endocrine glands are: the thymus (only in youth), the thyroid, the liver, the mouth, the stomach, the pancreas, the intestine, the spleen, the suprarenals (adjoining the kidneys), the appendix (adjoining the caecum); the testicles or ovaries, according to sex; and in the brain, the pineal, and the hypophysis (pituitary). There are also the secreting glands which provide the elements necessary to the life of certain of our organs, in particular, those of sight, hearing, smell and taste.

The Omnilateral Action of the Hormones

The hormones act on the secreting glands in the following ways:

1. They promote growth by the normal secretions of the thymus and thyroid.
2. They promote the proper functioning of the vasomotor nerves by secretions of adrenalin from the suprarenal glands.
3. They promote perfect digestion by secreting the digestive juices of the mouth and stomach (ptyalin and pepsin). In the duodenum, they secrete the juices elaborated in and discharged into this organ through the pancreas (pancreatins) and liver (bile).
4. They provide protection against intoxications (particularly those affecting the nervous system, where they cause paralysis), the hormones intervening to secure proper functioning of the white corpuscles (leucocytes, the killers of microbes) through the lymphatic glands and through the spleen.

Coordination and Adjustment between All our Organs

The hormones also have the function of coordinating and regulating the whole system of the endocrine glands, including ocular, auricular and nasal glands. They also have an action on the vermiform appendix and the glands of generation. The hormones have functions of extreme importance to our existence and health, as they coordinate and regulate all our organs, wherever situated.

Their functions are all the more remarkable in that they coordinate and regulate the mutual working of the endocrine glands, sympathetically arranged in groups of two.

Removal of the Endocrine Glands

The atrophy (or the removal) of an endocrine gland leads to the hypertrophy of the other endocrine gland with which the non-functioning gland was linked sympathetically. In women, the removal of the ovaries causes precocious senility and deterioration of character, with a resulting hypertrophy of the thyroid (which develops abnormally), which in turn leads to goiter. In men, removal of the testicles modifies character and tastes in a very characteristic manner, and at the same time causes fatty hypertrophy through the abnormal functioning of the other coordinated glands. In both men and women, the removal of a still healthy appendix is the cause of serious mental disorders. This is a serious matter and should make certain practitioners, whose bistoury is always ready to cut out the suspected organ, more circumspect in the future. Nearly always its painful inflammation (warning sign) is simply due to mistakes in diet, which have merely to be rectified for the appendix to heal rapidly.

Internal Hormones and External Vitamins

The hormones, which are purely hypothetical bodies of internal origin, are really only effects, as is everything which we need for life (but effects whose repercussions extend to every organ of our being). They are themselves generated by the vitamins, whose sole source is natural foods, all of them of outside origin. In other words, the active effects of the hormones are drawn from edible substances, which alone are able and intended to make complete provision of everything required for the conservation of balanced vitality in the human body and its component organs.

Their Common Origin: the Vital Radiation

The phenomena which used to be attributed to these hypothetical bodies, vitamins or hormones, have no real existence and are simply effects produced by the particular quality of each kind of living food. In other words, vitamins and hormones are only the product of the vital vibration made by the living food when it disintegrates, a vibration which extends to the organ with which each kind of food is in radiant accord.

Internal and External Balance between the Vital Vibrations and the Organs

By a rational choice of foods in which every kind of alimentary vitamin (really alimentary vibrations) is represented, we can be sure that each of the organs of the human body will function with the maximum of power and in harmony with its fellows, because it will have received the quantity of radiant vitamins necessary for its satisfactory functioning. Correlatively, these organs, while coordinating and regulating their physiological actions, will also live harmoniously and not disturb the reciprocal part played by the glands of internal secretion or make them secrete abnormally, excessively or insufficiently, as the case may be. Without such harmony, physical or moral health is impossible, since it is entirely dependent upon the well-balanced and correlated functioning of the glands. Everything that quickens and maintains our corporal structure must be external in origin. Organic life must be sustained by the absorption of the vibrations contained in living foods and not internally, as the false theory of hormones secreted by the endocrine glands themselves, demands. Only thus can it secure the effects of stimulation of its vital equilibrium. Organic life achieves this dual purpose in the following manner:

Firstly, the digestive organs (most of which are organs of internal secretion) will find in alimentary substances, which are still living, all the excitations in the form of inductive vibrations required for their perfect functioning.

Secondly, the other organs (these, too, all of internal secretion), both those completing the digestive apparatus, such as the liver, lungs and spleen, and all the other endocrine glands entrusted with ensuring in different ways the sum total of the phenomena necessary for the support of physiological life, will find in the intestinal chyme elaborated by the digestive apparatus, in the blood aerated by the lungs, and in the subtle essences of foods (really their aromas), the kind and nature of vibratory excitation most suitable for assuring perfect functioning of all our organs. Each of these digestive or other organs automatically finds the vibration whose wave-length harmonizes with it, selecting the vibro-excitative alimentary entity appropriate to it according to its anatomical conformation and biological function.

That is to say, the vibratory flux (the vitamins), which is of alimentary origin and automatically selected to vibrate in harmony with one of these organs, can activate only that organ and no other.

Emission and Reception of Alimentary, Radiant Waves

The alimentary radiant waves behave in the same way as the electromagnetic waves in radio transmission, which can only be selected, i.e. received, by a receiving apparatus (comparable to a human organ), so arranged that the number of its coils is equal to that of the organ of the transmitting apparatus (the vibratory vitamin situated in the food).

The food emitting radiations harmonizes, according to the number of radiations emitted, with different organs of the human body which are only the receptors produced by these radiations.

Classification of the Alimentary Radiations

If we regard vibratory food as a broadcasting station emitting waves, what happens can be expressed by saying that the organ which it has to excite in order to make it function (the wave-receiving set) is adjusted to receive the particular vibrations of the foods appointed for it, and no others. This is exactly what happens in practice. While they are still living, the foods can be regarded as endowed with a particular vibration, whose number of waves is larger or smaller according to the class to which they belong. We can adopt for the classification of the alimentary waves, the same classification as was made for the vitamins, using it as a basis and following the same order, provided that the word "vitamin" as it stands be replaced by the word "vibration" and intelligence and truth thus satisfied.

First of all we must separate the essentially digestive organs from those whose functions extend to every part of the body. If all our organs are regarded as being capable of animation and function only when a certain number of vibrations reaches them to set them working, the vibratory action which will be produced (the direct effect of the living food upon the digestive organs and the emanation (entity) of the vibratory action upon the distant organs) will intervene as an excito-induction on each of the organs with which its wave length is in harmony. But it will only affect that organ and no other; for our organs can only function if they receive an excito-induction produced by a wave length equal to that of the animating vibration.

The foregoing explains the reason (applying this theory to the theory of the effects of vitamins) why one food can only act upon one organ, while another food only has and only can have vibratory action on some other particular organ and not on all the other organs; hence the necessity of varying these foods and of taking certain kinds which comprise nearly all the vibrations. Vitamins, like hormones, are not causes with fixed effects, but are merely the result of the vibratory effects caused by the excito-induction which is generated by the various kinds of living foods.

This explanation provides a concrete solution, in agreement with existing knowledge, to the somewhat abstract problem of vitamins and also hormones. Vitamins are not existing bodies capable of classification and definition, but are really only the consequence to our organs of the harmonious effects caused by the vibrations of the alimentary molecule after its integration in our intestines and its transmutation in our organism.

These vibrations whose source is to be found exclusively in the molecule and its alimentary entity, vary both in number and width or range, according to the kind of food from which they emanate.

They may be classified as follows:

(a) The vibration of the molecule of the foods where we find the vital principles attributed to Vitamin A (unsaturated fats) will differ in number and breadth of waves (red waves) from

(b) the vibration of the molecule of the foods where we find the vital principles attributed to Vitamin B (wholegrain cereals, eggs and raw milk), which will differ in number and breadth of waves (yellow waves) from

(c) the vibration of the molecule of the foods where we find the vital principles attributed to Vitamin C (fresh, unsprayed fruits and vegetables), which will in turn differ from the two preceding vibrations in number and breadth of waves (blue waves).

Radiant Harmony between our Organs and Foods

We have explained the reason for the vibratory harmony which exists between each of the organs of the human body and each of the waves emanated by the different kinds of foods necessary to secure the total nutrition of the whole organism. We are now able, with this concept, to explain the cause of every organic atrophy, such as the cessation of certain digestive functions (post-digestive disturbance or abnormal elimination) by the single fact of the lack of certain kinds of living foods.

The excito-induction which is indispensable for the activation and properly balanced functioning of every organ of the human body, and thus for its optimum health and harmony, can have its source only in the products liberated in every part of the body after the disintegration of the substance and essence (vibratory aroma) of living food.

"Man does not die; he kills himself," says Jean Finot, who wrote *La Philosophie de la Longevité;* and Seneca, Stahl and Broussais said the same. Human life is something more than the sum of the mechanical forces and chemical combinations at work in our body. Above and beyond matter there is a superior force directing and influencing the former, an essential principle, a vital dynamism, variously called at different times and by different philosophers, Soul, Thought, Reason. This must be our guiding principle in the regulation of our physiological functions. To obtain health we must will it. It is a truth, for which there is more than sufficient evidence, that a sudden outburst of passion in a hitherto calm and balanced individual, is enough to sow the seeds of disease. Grief or fear can have a considerable effect on the system. Hope and joy bring the dying back to life; they warm and rejuvenate the blood and re-invigorate the muscles. We see, then, that thought can affect the body. We have within us the fountain of youth which can make us perennially young and strong, enabling us to reach, without infirmity or decline, the limit of life assigned to us by nature. We think it extraordinary to live to be ninety. We ought to live to be more than two hundred.

Many years ago I made a note of the charming thought of an English writer, Mrs. Nora Wynne. She says: "Age is a voluntary evil which is not at all inevitable. It is not a question of dates and birthdays. It is a matter of natural tendencies and dispositions. The age of the body is irrelevant so long as the intelligence, the emotions and the instincts remain young. Youth is not a time of life; it is a quality, a trait of character, a mental condition. Old age is nothing more than a form of ignorance and egotism."

What must we do to keep this precious trait to an advanced age, this precious state of mind? We must not only live abstemiously, and observe the laws of health; we must also practice mental hygiene. Keep a watch on your thoughts, brush aside all feeling of depression and discouragement and close the door to doubt, morbid fancies and every kind of phobia on the question of age and illness which could disturb your peace of mind.

At the risk of quoting other people too much, I am going to call an expert witness to this fact. Listen to the admirable evidence of Xavier Pelletier, who has specialized in the problem of solution to these universal questions: "The discipline of the inner life, able by progressive effort to brush aside the passions which bring about man's downfall, this calm philosophy that teaches us to master our impulses, the violent upheavals of our soul, has been known throughout the ages. We find it among the Indian Theosophists; it was the basis of Stoic morals. Seneca gave good advice to the neurasthenics of Rome when he wrote to his friend Lucullus: 'Pain is easy to bear when it is not exaggerated by dwelling on it. Thinking makes it so.'

"Any kind of psychological disturbance means lowered resistance. Though we do not die of grief or fear, they affect our system in such a way that it easily falls prey to the enemy at the gate. The depressing effects of physical conditions, or climate, and daily contact with our fellow-men must be combated by an unshakable optimism. We succeed in this by cultivating the will. Not an imperative will implying effort, but a reasoned, permanent will which has become an instinctive, reflex act. We must master life and dispassionately control our actions. This is the way in which we can, without being involved in a sterile egoism, ease our great sorrows, suppress anger, eliminate from hatred all but the element of indignation that drives us to action, and know the joy of forgetting and forgiving."

We could not, if we tried, express this better.

LONGEVITY AND THE MIND

This peace of mind and soul will be a sure shield against disease and old age.

Buffon has said that what destroys balance in a person is an attitude of mind which dwells on evil at the expense of good. If you are not afraid of death, death will retreat before you. Most people, when they reach a certain age, think that they have come to the end of their lives. Just a few sad years of the discomfort and decrepitude inseparable from old age, and that will be the end. Their minds are filled with gloomy forebodings, their horizon is narrowed. They lose their buoyancy and are haunted by the approach of death. They are victims of the worst sort of auto-suggestion and cannot react. In their distress they eat badly; nutrition is impaired, their sleep disturbed, and their poor nerves, frayed by all these harmful thoughts, upset the whole system. The least symptom is seized on by their disordered imagination. They brood on it until it develops into the disease which will ultimately kill them.

Numerous scientists share this view. In the eighteenth century Stahl, in his doctrine of Vitalism, asserts that apart from violent unforeseeable death, death is always "real suicide or rather murder committed by the soul upon the body through ignorance or folly." We must defend ourselves against evil thoughts which shorten life, and we must learn to keep in good spirits, to treat death with contempt, and to cultivate the calmness of spirit which breeds centenarians.

THE TECHNIQUE OF PSYCHOVITALITY

"Happiness is not easily found. It is difficult to find it within ourselves and impossible to find elsewhere." (Chamfort)

THE BEST WAY OF IMPROVING THE INDIVIDUAL

Our aim is the *omnilateral vitalization* of our psychophysical organism, by the absorption of all the vital energies in nature, by the elimination of all morbid and inferior substances, by the improvement of all our existing senses and psychophysical capacities, by securing individual optimal perfection and maximal longevity through continued optimal and omnilateral evolution towards perfection based on existing preconditions, and by the creation of successive phases of progressive superiority:

1. by continual contact with the natural forces (sun, air, water, earth, higher cosmic radiations);
2. by systematic exercises and breathing pure air (through the skin and lungs);
3. by an optimal unfired (raw) diet, qualitatively (wholesome, simple combinations), quantitatively (small amounts), and modally (complete mastication);
4. by regenerative fasting;
5. by omnilateral concentration of thought directed optimally and continually to every problem, through the creation of progressively higher mental currents from the point of view of ethics and knowledge, and the attraction and absorption of all the higher currents in increasingly higher affinity;
6. by the optimal realization of all these thoughts through omnilateral activity in the sphere of ethics and knowledge.

SELF-EDUCATION AND RE-EDUCATION

"Everyone receives two kinds of education: the one given us by someone else, and the other far more important education which we give to ourselves." For success you must on the one hand govern people and circumstances, while on the other you must first of all learn to govern yourself. Self-mastery can be acquired like an art or a sport, drawing or tennis. If you wish to control your destiny, you must learn to educate yourself.

Body and Mind

The body is the machine of the mind; it must be in good working order, or else there is no possibility of success. "Bodily vigor must be maintained to preserve that of the mind." (Vauvenargues)

CONFIDENCE AND OPTIMISM

Have Confidence. Physical, intellectual and moral counsel can only be attended with success on one condition: faith in the result. The brain, consciously or unconsciously, governs all the organs of the body: the movements of the heart, the working of the lungs and digestive tract, nutrition in general, all are consciously or unconsciously dependent on the mind. Consequently, each time we get discouraged and no longer believe in good health, each time we complain of some trouble, our cerebrospinal system gives the organism harmful directions, causing illness and even death. Optimism creates healthy glandular secretions, clean blood and regular circulation. Pessimism creates poor glandular secretions, intoxicated blood, and irregular circulation. So never talk about your bad health, do not worry about your physical troubles, do not say you are suffering. Be assured that a comprehensive natural way of life will give you good health. If you apply these principles you will put yourself on the right road to a complete cure. Your cerebrospinal system, through your faith in success, will order your bodily functions to behave normally.

Ourselves the Cause of our Misfortunes

Most people complain of their bad luck, their diseases, their accidents, the tiresome occurrences thwarting the realization of their desires. This means they are dominated by events instead of being their master, or turning them to account.

An American laboratory of experimental psychology has shown that out of a hundred people adopting a career, only three are successful. The remaining ninety-seven lead a poor, joyless life far different from what they had a right to expect. Out of this host of unfortunates, there are one or two who owe their failure to bad luck. The rest have only their personal defects to blame.

CHIEF DEFECTS

The chief defects which harm health, success and happiness

are numerous, but four predominate: emotionalism, fear, hatred and pessimism.

1. *Emotionalism.* "People who have the misfortune to be accustomed to violent forms of pleasure lose their taste for simple ones and always find themselves bored in their troubled search for joy." (Fénélon) From a physical point of view, a state of hypersensitivity leads to troubles in connection with the heart and circulatory system. With faulty nutrition, it contributes to premature arteriosclerosis by overworking the heart and blood vessels. The brain suffers indirectly. From a moral point of view, emotionalism, pain and anxiety, cause discharge of nervous influx, the chronic source of timidity, lack of will, depression and discouragement.

2. *Fear, timidity and anxiety* are the children of emotionalism. These parasitic defects absorb energy, destroy the will, and paralyze useful activity. Fear destroys our mental faculties. What is more, it impairs our health by organic poisoning of our system (harmful glandular secretions). *We should only be afraid of one thing: of being afraid.*

3. *Hatred* and its derivatives: envy, jealousy, anger, etc., produce poisons (through disturbance of glandular secretions) which affect our health and use up a large part of our energy and vitality which are so essential to thought and action.

4. *Pessimism* systematically sees the bad side of people and things. Pessimists see their superiors as imbecile and unjust, their colleagues as shifty intriguers, and their subordinates both lazy and dishonest.

Pessimism and its derivatives, bad temper, surliness, a critical outlook and ill-nature, spring from insufficient self-esteem, from absurd vanity, weak will and from certain organic troubles.

The pessimist is out of tune with things. There is something essentially wrong between the pessimist and his environment. He would like the world to harmonize itself with him and his personality, while really it is he who depends on the world and should fit himself into the general prevailing harmony.

The philosopher Young rightly declared that one's greatest riches lie not in good fortune, but in a happy disposition.

THE CONQUEST OF HAPPINESS

In order to establish physical health, it is necessary to suppress unnatural habits and adopt the hygienic measures which the organism requires. The next thing is to await with confidence the inevitably successful result. *In the same way, one must, for the attainment of happiness and success, establish good conditions of physical, mental and moral hygiene and await the desired result with the certainty that it will be attained.*

You ask what you must do to keep these rules.

1. You must train yourself to be calm.
2. You must adopt an optimistic and benevolent attitude.
3. You must educate your will.
4. You must engage in some useful occupation.

1. *Training in Calmness*

You impressionable, nervous subjects, you must be convinced that there is no possibility of health or success unless you overcome your hypersensitivity. You must learn not to let yourselves be put out by any accident befalling you, not to be depressed by any vexation or preyed upon by anxiety.

How is this transformation to be brought about?

Avoiding Alimentary, Physical and Nervous Excitation

(a) By avoiding every cause of over-stimulation in your diet: alcohol, coffee, tea, anything containing caffeine, cocoa, chocolate, sugar, all flesh foods, all junk food, all chemicals, drugs, etc.

(b) By avoiding the causes of physical over-stimulation: noise (especially traffic and amplified sound), over-use of television, time spent with people who smoke, unintelligent use of evening hours, etc.

(c) By avoiding the causes of nervous or emotional over-stimulation. No more "junk" plays, films, books or television; no dealings with nervy, agitated or anxious friends. Instead, seek the company of those who are calm, balanced, and serene. Suppress all signs of agitation, gestures of impatience, recriminations (of self and others), hasty steps, hurried eating, preoccupied appearance. Control every "parasitic" movement. Do not drum on the table with your fingers, as a sign of impatience, nor cross your legs when seated. Stop biting your lips or chewing your nails. These pointless acts enervate and tire you.

Autosuggestion

(d) By practicing *autosuggestion.* This is how it is done. You set down in writing and fix firmly in your head one or more formulae, expressing the quality you wish to acquire, and you learn them by heart. For example, suppose you are a lecturer who wants to be cured of nervousness which prevents your speaking in public. You compose a formula, such as, "I have confidence in myself. My intelligence is increasing, and I speak better every day." Then, each morning and evening you go to some room by yourself and lie down, in absolute quiet. Close your eyes and repeat aloud or half-aloud the formula you have memorized. You repeat it twenty or thirty times. Even if you do not have faith in it, the idea penetrates the subconscious. Then you resume your normal occupation. Your mind is impregnated with the idea. You repeat these acts and formulae for a week, two weeks, or a month. You will be surprised at the results you will obtain. As little by little success is sure to greet your efforts, you will gain self-confidence, more and more as you realize the progress you are making.

Autosuggestion is not only useful in the acquisition of qualities of character. It is also one of the conditions for regaining health. The cerebrospinal system acts at one and the same time on the psychological functions, on voluntary acts and organic functions. It is the brain which tells the stomach to break up food, the heart to beat, and the lungs to breathe.

Each time you give yourself an optimistic autosuggestion favorable to health, such as "I am feeling well," your bodily functions will work better and you will hasten the trend towards cure and optimal health.

Have you ever tried, since you grew up, to learn a sport or an art or a skill? *You have succeeded in your efforts by repetition, by interest and habit and by faith in your ultimate success.* If you wish to learn a skill you must be sure you will succeed. You must count on success from the start and await it confidently. Very well! You can also learn to be calm, attentive, energetic, just as you learn the violin or a foreign language. All that is needed are training exercises to form new habits. If you want to be calm, force yourself to bear a calm, patient expression. Go into an empty room three or four times a day, lie down, and without moving so much as a finger, repeat thirty times aloud or half-aloud the phrase, "I am calm." Eventual success is virtually guaranteed.

2. *Adoption of an Optimistic Attitude*

We should adopt a physical, mental or emotional attitude as we adopt a style of house, a fashion in clothes, a color of paint. If you are anxious to transform yourself, you must think continually of the ideal you would like to attain, keeping the image constantly before you. Let the new "I" be optimistic, that is, good-humored and satisfied with everything. Let the new "I" be kind, polite, conciliatory, lovable. Let the new "I" make others happy, and personal happiness will surely follow. The pessimist demands that everyone should fit in with his scheme of things. The optimist finds it simpler and more effective to put himself in tune with everyone and everything else.

Optimism helps the organs to function as they should, while pessimism floods the body with poisons, predisposing it to disease and premature old age. Pay no attention to what is ugly. Avoid saying anything sarcastic or derogatory. Never give pain to others, nor to animals, birds, plants, or trees. "Everything lives, and everything is holy." (Ady) Avoid contact with people who like to grumble and who see the bad side of everything, until you have yourself acquired the habit of optimism. Train yourself to be just as kind to your family as you are to your friends, acquaintances, or co-workers. You will make others happy by being happy yourself, and you will then know true joy. "Politeness," said Montaigne, "costs nothing and buys everything, for good manners are the ornament of action."

The Radiant Magnetism of Altruism

The kind-hearted person unconsciously captivates people and wins their hearts. The eyes, voice, gestures, the whole personality radiates charm. There is a radiant magnetism around the kind-hearted person which grips you. The more you develop kindness and optimism in yourself, the faster you will see the disappearance of care, grief, and physical discomfort; your thinking will become clearer and your wishes more fruitful, as you find others collaborating with you. If you want to acquire the habit of seeing the bright side of people and things, you must keep a watch on yourself to start with. Repetition will develop the new attitude of mind. *Repetition of an act creates a habit. The habit, as first as fragile as a spider's web, ends by being as strong as a steel cable. Goodwill and optimism bring success in their train. The most productive work is work coming from the hands and brain of someone with joy in his heart.*

3. *Education of the Will*

Of all the faculties of use in your physical and moral re-education, the most useful is the will. The doing of a voluntary act consists in the transformation of an idea into action, energy and continuity of voluntary efforts towards a determined goal.

A person is endowed with will who accomplishes everything one has decided to achieve, who does not let one's decision be changed by people or things. The immense, unfailing power of the will has been continually affirmed throughout history. *Vouloir, c'est pouvoir.* "A person," said Thomas Buxton, "can become anything he wishes provided he makes up his mind firmly and sticks to it." In his book called *To My Son,* Paul Doumer writes: "In the struggle for existence, the man of energetic will succeeds better than any other, but his will must be exercised without rest or repose and must be directed to his own improvement and to every act of his life."

The will permits us to realize every thought elaborated by the mind, but that is not of itself sufficient. The thought must be a good and useful one, and for that we must conceive it only after due and careful reflection. The will must be intelligent, and intelligence is developed by concentration.

Concentration is the quality of mind which allows it to make all its faculties converge on a single object to the exclusion of every foreign thought, every distraction. The concentrated mind is directed exclusively to one thought, one feeling, one subject, or one object at a time, and so exercises great power. By its means, powers of judgment, deduction and memory are developed. The student capable of working in a library without noticing the comings and goings of other readers, as if he were reading in solitude, has the gift of concentration. The *efficient* practical person who *achieves* is the person who knows how to concentrate.

4. *Useful Occupation*

It is not enough for success to have good moral qualities and good health. A further requirement is that these qualities should be applied to something useful. Success comes in direct proportion to the service rendered to others. It is impossible to be useful to humanity without benefiting oneself. Choose an occupation responding to a need. As the needs of the human race are unlimited, so are the opportunities for success. The choice of a useful occupation once made, one must cherish and devote the best of oneself to it. The day you begin to apply yourself to something of real utility and attain true competence, that is to say, *excellence* through repetition, habit and perseverance, you are sure of meeting with great success, because the world cannot do without you. This is a mathematical truth.

ATTAINING EXCELLENCE

Ralph Waldo Emerson, the American philosopher, once said, "If a man write a better book, preach a better sermon, or make a better mouse-trap than his neighbor, tho' he build his house in the woods, the world will make a beaten path to his door."

You are worth many, many times more than you are at present. You are hiding under a bushel precious ability, the fertile seeds of health and success. Make them bear fruit by re-educating yourself. Is this the path that leads to happiness? Yes, but to enjoy that happiness to the full and to keep it, it must be shared with others. Help others to gain happiness for themselves; shed the light of happiness on others, and it will shine all the more on you.

Realize that your life has no value except in action and that your actions will be all the more effective if they contribute to the well-being and joy of your fellow creatures. Only then will you be able to say that you have won success and happiness. There is no true happiness save one given or shared. You are yourself the architect of your health, will, success and happiness. There are some who have the good fortune to possess health, will, intelligence or the social qualities which attract and influence others, by their heredity or education. There is certainly "good luck" in having spontaneous possession of these qualities, but it is only where conditions of happiness dating from infancy or early childhood are found that this "good luck" exists. And anyone can acquire this "good luck" and have all these capacities simply by re-educating themselves. They have only *to will them intensely* through the practice of autosuggestion.

Self-Reformation

Reform of oneself takes some years, but from the very first you get appreciable and encouraging results. Have you ever watched acrobats performing feats at the circus which demand not only skill and agility, but great flexibility as well? Some years, perhaps only months earlier, they were as lacking in acrobatic skills as you. Little by little, at the cost of long patience, great perseverance and energy, they mastered the feats you are watching with admiration. It is not much harder to make you an energetic, calm, lovable, influential being than to make you an acrobat or a pianist. A simple, natural, healthy life, exercises in physical and moral culture, patience, energy, the practice of autosuggestion, these are the conditions for the possession of health, achievement of one's goals, success in one's chosen work, or happiness itself.

Program and Means

1. *Idea* is the clear vision of the end to be attained and the best program to follow in order to arrive there. To sustain the clear vision of the idea, the following capacities are required:
   - (a) *Concentration,* which consists of three psychological stages: attention, self-control, and continuity.
   - (b) *Suggestion:* autosuggestion, study of the conscious, the unconscious, the subconscious, and the superconscious.
   - (c) *Memory.* One must develop a good memory for numbers, names, objects, events, and people.
2. *Will* is the execution of the *idea.* It is the passage from an idea to an act. Study the means of educating and developing the will.
   - (a) *Physical Means.* Adopt a positive attitude, indicative of strength. Suppress parasitic movements. Maintain a forceful, determined bearing. Do some form of physical exercise for half an hour each day with care and attention. Observe general and alimentary hygiene.
   - (b) *Moral Means.* Self-control, self-mastery, mastery of one's words, thoughts, feelings, acts. Train yourself to be calm. Avoid nervous people. Try always to do everything to the very best of your ability. Have perseverance. Practice continuity and be centripetal: never stop half-way in the performance of a task you have undertaken.

Driving Forces of the Will: Faith, Hope, and Love.

Faith is confidence in oneself, confidence in human power guaranteed by heredity and evolution. Bear in mind the example of animals and plants which, by breeding and selection, slowly lose their imperfections and approach a type of perfect beauty. Suppress and control all feelings of timidity, doubt, anxiety. At first, apply yourself only to tasks which you can master; gradually undertake more difficult ones; little by little, you will come closer and closer to perfection.

Hope. Have unquenchable optimism in the face of all and in spite of everything. Shut your eyes to sad, ugly things, to the imagined defects of others. Maintain harmony in and around you. Do not let the events of the world or the actions of others disturb your equilibrium. Do not criticize. See only what is good, whether in people, things, or events.

Charity. Be kind to all forms of life. Have not only good will, but also enthusiasm—the sacred fire. Suppress for ever in your ego the traits of envy, jealousy, hatred, bitterness and spite. These are harmful parasites and superfluous luxuries, exhausting your vitality, causing nervous discharges, creating poisons in your glandular activity, destroying your health.

3. *Order* is the reflection of the cosmos, where everything has a place and a time in the great universal methodology.

(a) *Order in Time.* Make a program for a month, a week, a day, to cover each of your tasks. Whatever you plan to do, be sure you carry it out to the best of your ability. Have order even in the smallest details. Every evening, try to keep a journal, before going to bed. Trace back in your memory the day just past, starting with the last act and finishing with the first. At the end of the week, draw up a "balance sheet" for the past seven days. Do the same at the end of each month. This is a good way to develop memory and form the habit of order and organization.

(b) *Order in Space.* There should be a place for everything, and everything should be in its place. Arrange your possessions as carefully and logically as if you were blind.

EDUCATION AND HUMAN SOCIETY

Human society is the greatest occupation of mankind, but without education all human values would remain in the hands of one generation only. Through education we ensure the continuity of priceless human values.

Constructive Work—Creation of Values
Destructive Work—Destruction of Values
Educative Work—Preservation of Values

Education is the means to bequeath to the present generation the experience of the past. Otherwise, each generation would have to start all over again. Therefore, education is a precondition of evolution and progress. How does education work in society and transmit values?

Nature is the basis of all our culture and civilization. Without nature, there would be no civilization, no possibility of life, no creation of man would be manifested. The path of life and of society is nature. The laws of nature should direct society, man, and all human activity.

In the structure of society, nature is the foundation. The material basis (labor and technique) rest on the foundation of nature. Then the division of producers and non-producers leads to the economic basis of society. So the substructure of society consists of material and economic bases. On this substructure rest the superstructure of state, bureaucracy, army, and political superstructure. Over this is the ideological superstructure: law, religion, morality, philosophy, art, and literature. The sciences belong to both the substructure and the superstructure.

In *space* there is the structure of values transmitted by education in *time*. The function and content of society are the same as the function and content of education. The values transmitted by education depend on the values possessed by society. The way in which education fulfills its task depends on the society in which the education exists. Both the quantity and quality of education depend on society, and also the mode of education, the method of its organization.

The aim of nature is harmony and balance. The true test is whether society is in harmony with nature, or not. Laws that are in harmony with nature should be kept, and all those not in harmony with nature should be destroyed. Even education is useless, whatever its method or quality, unless society is in harmony with the laws of nature. If society is not in harmony with the laws of nature, we must replace the present education with one that is in harmony with those laws.

Society must adapt itself to nature, and not nature to society. Societies appear and disappear, but nature and its laws are eternal. The laws of society are always changing: there was one set of laws in the age of slavery, another set in the feudal age, still another set in the age of capitalism. But the law of gravity, the law of the association and dissociation of matter, the speed of light, the appearance and disappearance of the solar systems, constantly remain the same. Societies are destroyed and disappear, but nature is eternal and never disappears. The laws of nature govern every other law. The natural laws are the directive laws, coming before human laws or the laws of society. For instance, a change in climate can completely destroy a society. But nothing—not parliament, nor any political party, nor any human force—can alter the functioning of the solar system of our planet. The laws of nature are the greatest force in existence. They aid everything that is in harmony with themselves and destroy everything that is out of harmony. This is why the individual must understand the laws of nature and adapt himself to them—this is equally true of society. True education is the knowledge of all the laws of nature and of the best way to transmit such knowledge to the next generation.

Does real education exist today? If there were real education, society would also be harmonious. Let us consider whether society and education at present are in harmony, examining the diagram of social forces.

WORK AND TECHNOLOGY

There are thousands of schools, institutes and universities in existence for giving instruction in labor and technology in all kinds of professions, but are their teachings in harmony with the laws of nature? Let us see what the laws of nature say about work.

Work is a precondition of existence, provided that it is healthy, but in our present society work is completely out of harmony with nature. So much of what is considered important work is not only superfluous and unnecessary, but completely out of touch with fresh air and sunshine. The rhythm of work is out of harmony with natural law. In harmony with nature, work can be one of the greatest pleasures of human life. But when we are compelled to work under conditions which are out of harmony with our natural disposition, the nervous system is destroyed. Work then becomes a source of disease and unhappiness. One half of the world is unhappy because it has inharmonious work to do, and the other half is unhappy because it has no work to do. At present, neither work nor education is in harmony with the laws of nature.

The situation is the same in regard to technology. What should be the real aim of technology? Ideally, machines should be doing all the monotonous, mechanical, unhealthy work that does not contribute to the development of man. In other words, technology should free mankind from all work which is unworthy of humanity, so that man can be free to do that work which is in harmony with the laws of nature, work which contributes to our individual evolution and that of the planet. Technology should exist for our benefit, not the other way around.

The laws of nature tell us that man is the measure of all things. Man is the aim or purpose of nature, because man alone of nature's creations can understand nature's laws. But in our present era, technology does not serve man. Instead, it is man who serves technology. It is clear that technology is not in harmony with the laws of nature, for instead of freeing man, it enslaves him. The rhythm of work is not in harmony with man, but with the machine which dictates to man. Therefore, man is the servant of the machine. Constant contact with technology, at the expense of our relationship with the forces of nature, can impair the human nervous system and destroy health. If the present trend continues, man will disappear and only machines will remain.

Present technology causes misery. Nature gives every good gift to man: sun, air, water, fresh foods, trees, grass, etc. A change must be made, and technology must be made to function on a natural basis. Only if technology can enhance a life in harmony with nature, and not threaten human health or the ecology of our planet, should it be allowed to continue its existence.

NATURE AND SOCIETY

The aim of nature is harmony and balance, but this is not possible with the inequalities and injustice in the present forms of society in the world today. In the human organism, just as in the solar systems and cosmic space, every part is there to help the totality, just as the heart and lungs help the entire organism. Mutual aid is a law of nature, both in cosmic space and in the human organism. But in the false divisions and warring societies in the world today, people try to enslave each other instead of helping one another. Egoism and selfishness on the part of those in political power result in ignorance and fear in those without power, conditions which are dangerous because they are contrary to the laws of nature. Instead of selfishness, we need reciprocal love; instead of ignorance, we need knowledge of the laws of nature. If a change is not brought about, nature will destroy present society and all those who do not live in harmony with its laws, just as it has done in the past. Egoism and ignorance can lead to war, and the kind of war made possible by our present technology would destroy all civilization and humanity itself. It is most important to replace our present ignorance with knowledge and to replace disharmony with harmony. Instead of teaching the importance of preserving the natural ecology of our planet, we teach better and more efficient methods of eliminating our trees and vegetation, ruthlessly and methodically drowning the planet in a morass of toxic pollution. Instead of teaching the knowledge of the laws of nature and the principles of active love, we teach hatred of other nations, races, religions and creeds.

NATURE AND GOVERNMENTS

Are governments, bureaucracies, armies, political systems, etc., in harmony with the laws of nature? Absolutely not. In spite of some beautifully-worded constitutions, most governments do not exist for the benefit of their constituents, but against other governments, supported by their (willing or unwilling) constituents. Bureaucracy creates artificial, inharmonious conditions, multiplying itself in a nightmare of red tape. Following the stone age and the bronze age, bureaucracy has created an age of paper. For the continuity of life, sun, air, water, trees, grass and precious top-soil are necessary, not paper. In this bureaucratic age, it is impossible to be born, to die, to travel, to buy or sell, without paper. Natural conditions are entirely replaced by artificial conditions.

Armies exist for the purpose of bringing death and not life to mankind. It is not a law of nature that we should kill one another. That is the role of wild animals who are nature's *means* to perfection, while man is not a means but the *aim* of nature. The aim must not be destroyed. The aim of nature is peace and harmony, not war and destruction.

Thousands of political systems and governments have existed, but they have never helped the progress of mankind. The work of political systems is the accumulation of mountains of words and the fabrication of false catchwords. Each political system declares all of the others wrong and itself right. Political systems, however, are incapable of healing, educating or constructing the life of mankind. This can only be done through a life based on the laws of nature.

NATURE AND RELIGION

When the ideological superstructure is based on an unnatural substructure, and not on the laws of nature, then ideologies and religions are not based on harmony and mutual love, but on the right of the stronger over the weaker. Religions should bring harmony and unity, but as it is, they bring separation and strife. There are more than eight hundred religions, all working against each other. While the underlying essence of all religions consists of harmony with the laws of nature and the cosmos, this original purity and simplicity has been lost in a forest of warring dogmas and theologies, until the existing religions would not be recognized by their founders: Buddha, Moses, Jesus, Zarathustra, etc. It is not the fault of the founders, for they represented higher values of mankind, and their original teachings will always represent the highest ethical standards of humanity. But the disciples do not have the spiritual and psychological maturity of the master; their teachings have been misunderstood and deformed, becoming more and more petrified as the years passed, their vital, original, living ideas being replaced by rigid dogmas and dead scriptures. The *true* spirit of vital life represented by these religions could invigorate and vivify society. Dead words and dogmas divide people and lead them against one another. Jesus said, "The spirit quickens; words kill."

NATURE AND MORALITY

There is one kind of universal morality which is based on the laws of nature, and another morality which is invented by society. There are as many of these latter kinds of morality as there are nationalities and social groups. What is moral in one place can be immoral in another. What is considered moral in one age can be considered immoral in another. The only real morality is active love, which is the same at all times and in all places, transcending artificial, man-invented moralities.

NATURE AND PHILOSOPHY

Up to now, philosophers have only interpreted the world in different ways—the important point is to change it. There are many hundreds of philosophies, but only one reality. All philosophical systems are unilateral, not being based on the totality of the laws of nature, but on one-sided laws representing this or that form of society or time of history. Only that philosophy which is based on the laws of nature can have universal validity.

NATURE AND THE ARTS

The function of all forms of art should be to beautify human life, but the vast majority of human beings do not know nor enjoy nor understand the arts. Lack of education and the present mediocre standards of society are mostly to blame. How many people truly understand the musical language of Beethoven, or the spirit of expressionism in certain great paintings? Just because there are so few human beings capable of understanding real creative art and music, society bows to the wishes of the majority and replaces the arts with inferior substitutes which the masses can understand. Time that could be spent with da Vinci, Michelangelo, or Mozart, is wasted on junk art, junk television, junk music. And time lost in this way can never be replaced.

NATURE AND LITERATURE

The mass of mankind today is still in an infantile state of development and cannot or will not try to understand the great masterpieces of literature. Even those who understand have neither the time nor the possibility of truly appreciating them, and so inferior substitutes are provided for the works of Goethe, Shakespeare, Victor Hugo, Tolstoy, Dostoevsky, etc. Even today when the great masterpieces of literature are easily obtainable in any book store in most countries of the world, the majority would rather read cheap novels and magazines, or watch television and not read at all, apparently wanting to avoid the challenge and stimulus offered to the mind only by a great work of literature.

NATURE AND EDUCATION

Not one of the branches of the ideological superstructure is in harmony with the laws of nature, including education. Our present system of education in the so-called civilized world can only impart false values; the extreme chaos existing at present is due to the fact that neither the forms of society nor those of education are in harmony with the laws of nature.

Existing education is false in two respects: first, because it is falsified by the past, governed by lifeless traditions and dead, static ideas which are not in harmony with active, dynamic life. The directive force in education should not be tradition, but the laws of nature; second, because it is falsified by the present, by the existing state of affairs. New falsehoods are created by society and piled on top of the old, which have been taken over and incorporated in our present society.

The role of education should be to transmit the true values of the past to the future, to transmit to the next generation those values which serve, strengthen and regenerate *life*. But education at the present time does not do this. One of the great tragedies of our age is that the content of our culture is static, accumulated in libraries, petrified in rockbound institutions. There is no relation of culture to real life. The only solution to save mankind is to transform that static form of existing culture into a true educational dynamism, to make education an integral part of a simple, natural creative life in harmony with natural and cosmic law. Quantitatively, this would bring education to all people everywhere, and qualitatively, static, superfluous knowledge would be replaced with practical, natural knowledge, that kind of knowledge which, because it comes from direct and living experience, leads to wisdom.

Just as an organism dies or becomes ill if it is disobedient to the laws of nature, so does society when it ignores natural law. The only way to solve the great chaos now existing in so many areas of life, is to study and apply in practice the totality of the laws of nature and the cosmos. In labor and technology, in political systems and governments, in economics and education, and in all the ideological forms of the superstructure: law, religion, morality, philosophy, art, literature, music—one must search for and realize omnilaterality—the best forms of life—the principle of the optimum, which is dictated by the eternal laws of nature. We must disseminate this great knowledge so that everyone may understand, before it is too late. Harmony and balance can be achieved only if *all* are happy, for the whole consists of the atoms composing it, and where there are unhappy people none can be happy. This should be the true direction and content of education.

MUSIC & LITERATURE

Ideological Superstructure: *Law, Ethics, Philosophy, Art*

Political Superstructure: Government — Military — Parties

*****SCIENTIFIC SUPERSTRUCTURE: THE SCIENCES*****

ECONOMIC BASIS—PRODUCTIVE FORCES: LABOR & TECHNIQUE

NATURE * * * * * * * THE NATURAL BASIS OF SOCIETY

The Pyramid of Culture.

THE NATURAL AND COSMIC LAWS

The cosmos is governed by the cosmic laws of the movement of radiations. When a direct radiation is changed into a whirling radiation (atom), we have matter. When a whirling radiation is changed into a direct radiation, we have force, or energy. This is the rhythm of the universe. The higher rays are finer and stronger and traverse the lower rays which are coarser and weaker. The lower rays of the universe are governed by universal gravitation whose centers are the centers of all planets coming into existence and disappearing, while the higher rays are governed by universal vitality whose center is the cosmic ocean of the higher radiations in infinite cosmic space, existing eternally. Individual birth is the conquest of universal death by universal life. Individual death is the conquest of universal life by universal death. This is the rhythm of the universe.

Our planet is a part, limited in space and time, of the infinite and eternal cosmos. Our planet is governed by natural law and by special forms of universal cosmic law. Life on the earth (the adaptation of organisms to their environment) has seven primary and innumerable derivative laws.

The seven primary laws are these:

1. *The law of temperature.*

Life on the earth can exist only between two limits of temperature (on the surface of our earth). Extreme heat (the interior of our earth) and extreme cold (external cosmic space) destroy life (in man, animals, and plants). A grain of wheat, planted at the temperature of life, germinates. A cooked or frozen grain of wheat will never germinate. Every organism should preserve its life-temperature and introduce life into itself (natural external temperature), natural internal nourishment (sprouted seeds and grains, raw fruits and vegetables), and not death (cooked and artificial foods, refined foods, flesh foods, etc.).

2. *The law of air (respiration).*

Without oxygen, there could be no life on earth. Respiration is the link between the organism and the cosmos. Where there is life, there is breath. Where there is breath, there is life. The perfection of life depends on the perfection of breathing. The cessation of breath is the cessation of life, and vice versa.

3. *The law of water (blood).*

Without water, there could be no life on the earth. The circulation of water in nature has the same role as the circulation of blood in the organism. All organisms—humans, animals, plants—consist for the most part of water and so do their foods. The perfection of the internal organism depends on the quality of the blood; the perfection of the life environment depends on that of water. One must have living water (the sun-irradiated water of sprouts, fruits, vegetables, unpolluted streams and rain) and not dead water (sunless or boiled water).

4. *The law of the sun (light).*

Without sun, there could be no life on the earth. The limits of the zones of life are the limits of the zones of the sun's rays (alike on the surface of the earth, in the oceans, and in the atmosphere). Without sun, there would not be the right temperature for life, neither for plants nor animals nor humans—not even for living air and water, as sunless air and water are dead. The source of life is the combination of air, water, and sun.

5. *The law of movement (activity).*

Without movement, there could be neither cosmic nor terrestrial life. The essence of the universe, of life, of all organisms is movement, activity. There is not a single motionless point in the universe. All organisms absorb external movements (passivity) and emit internal movements (activity), observing the external natural laws and their own internal laws of movement. Predominance of activity is life. predominance of passivity is death.

6. *The law of thought (wisdom).*

The finest and strongest currents are the currents of the higher thoughts. Inferior organisms emit inferior weak radiations which cannot traverse the sphere of gravitation of our earth. But the superior organisms of the dominant races (on our earth, man) of all the planets of the universe radiate higher rays of thought whose currents overcome the gravitation of their planets and, passing through the sphere of gravitation, reach infinite cosmic space where they unite with the eternal cosmic ocean of all higher radiations of all planets: this higher cosmic ocean governs cosmic life and evolution everywhere and always, eternally perfecting and being perfected. The degree of perfection of every organism depends on the degree of superiority of its thought-currents, sent to or received from this higher cosmic ocean of life, which is a source of knowledge, strength, harmony and perfection, giving a new sense of "omnilateral orientation" to all higher organisms of all planets of the universe, absorbing from them everything higher and eternal. The lower and dead remainder decomposes and returns to the reign of planetary (earthly) gravitation (death).

7. *The law of pan-harmony (love).*

The universe exists thanks to the seventh and most fundamental law of pan-harmony (love), according to which all finite parts of the infinite universe are in harmony, (1) with themselves, and (2), with the totality. Everything exists for itself, for the other parts and for the totality, governed by the constantly created and perfected eternal cosmic ocean of love of the higher radiations.

Therefore, the cosmos, universal and terrestrial life is created and constantly sustained by the cosmic ocean of love. Similarly, the human organism is created by love (moment of conception) and sustained by the mutual love of all its parts, living and existing for the good of the others: brain, heart, hands, lungs, stomach, etc. The moment this mutual aid ceases, human life ceases, too. The law of love is simultaneously cosmic, terrestrial and individual: it is the most fundamental of all laws. Only those currents of thought which are governed (created, helped and attracted) by the higher cosmic ocean of love, are capable of overcoming the gravitation of the earth. All other radiations or currents of thought without love are inferior, dead, like air or water without sun. Higher thought-currents of love create and sustain harmony, equilibrium, health, perfection; loveless, inferior, ignorant currents create disease, disequilibrium, disharmony, and death.

Of these seven fundamental laws, the first four (temperature, air, water, sun) are natural laws governing earthly life, and the last three (activity, wisdom, love) are natural and cosmic laws governing cosmic, or universal, life. The first four laws are the precondition of earthly, human life, while the last three are the preconditions of cosmic, eternal life.

Earthly life is the mother and cosmic life is the father of man. This is the meaning of the words, "honor thy father and thy mother, that thy days may be long."

The more the cosmic and earthly forms of life and existence are in harmony with the totality of the cosmic and natural laws, the higher they are. The more they are out of harmony with these laws, the lower they are.

THE LIFE OF THE COSMOS

DYNAMISM OR UNIVERSAL KINETISM

There is only movement. Things do not exist statically; they are formed, broken up, transformed. It is the partial formations of universal movement which undergo transformation.

MATTER AND THOUGHT

The most elementary dynamic formations are those of matter (matter being merely an aggregate of the movement of atoms), *while the highest dynamic formations are those of currents of thought.*

THE MATERIAL PRECONDITIONS OF COSMIC EVOLUTION

The infinite variety of the dynamic formations of matter necessarily creates the preconditions of evolution toward the higher dynamic forms (the formation of the solar systems, of the planets, of the satellites, etc., the successive cooling of the planetary surfaces, of the atmosphere (air), of the seas (water) and of the solid layers (earth), and the progressive changes in the temperature (solar energy) necessary for the formation of organic life.

THE ACTION OF INORGANIC GRAVITATION

The lower dynamic forms of matter are governed by gravitation (their mutual attractions and repulsions). Gravitation tends to blend together all the dynamic formations of matter and draw them towards the center of each planet.

CURRENT OF THE HIGHER COSMIC RADIATIONS: ORGANISMS AND INDIVIDUAL CONSCIOUSNESSES

On the other hand, where the preconditions of the highest dynamic formations of matter are produced, the currents of the higher cosmic radiations (which are the results of earlier but practically identical evolutions) tend to group the lower dyanmic formations (primitive organisms, the beginnings of and changes in the infinitely various forms of individual consciousness) around centers, just as on a huge scale the suns, planets and satellites are grouped around their centers.

THE ACTION OF ORGANIC GRAVITATION (VITALITY)

The directive force of these dynamic formations (primitive organisms, etc.) grouped around a center is *organic gravitation* (vitality) generated by the higher vital cosmic currents according as the preconditions for their formations exist.

UNIVERSAL AND CONTINUAL STRUGGLE BETWEEN GRAVITATION AND VITALITY

This vitality is the opposite of inorganic gravitation. While gravitation draws the inorganic groups toward the center of each planet, the vital force pushes the organic cellular groups in the opposite direction, away from the planetary centers toward increasingly remote spheres.

ITS RESULT IS LIFE IN ENDLESS EVOLUTIVE PHASES

In short, life since the time of the most primitive dynamic formations, is simply composed of the resultants of these correlative forces, gravity and vitality, developed through endless evolutive phases.

THE VICTORY OF GRAVITATION IS DEATH

Every organism whose vital force is overcome by gravitation breaks up into its inorganic elements and then becomes once more subject solely to gravitation (reign of death).

REPRODUCTION, CONTINUITY, EVOLUTION AND LIBERATION OF LIFE

However, the vital force, with each new generation, gives rise to higher and higher manifestations of life in proportion to environmental conditions and the cosmic forces present permit of such manifestations (vegetation, animal kingdom, man). In these manifestations the effect of gravitation progressively decreases, while vital force increases more and more.

The higher the material formation of life, the greater its ability to establish and produce in itself the cosmic vibrations of a higher type.

LAW OF THE DEGREES OF COSMIC SUPERIORITY

The degree of superiority of the material formations of life is directly proportional (a) to their ability to establish the cosmic vibrations of a higher type in such life, (b) to the faculty of producing such higher mental currents, and (c) to the degree of development reached by such mental currents.

LIFE ON OUR PLANET AND THE OTHERS: MAN

On our earth, the most developed form of life is at the moment represented by man. On planets older than ours incomparably higher forms of life succeed one another in infinite variety.

THE INFINITE–ETERNITY–COSMIC PURPOSE

Just as the cosmos and its life are unlimited, so is the changing variety of these dynamic formations without end. Consequently the aim of the infinite can only be infinite, since there is no objective already attained which cannot be transcended with the help of time and the presence of the necessary preconditions. The infinite ahead of us is as eternal as the infinite behind us. If a final, universal objective were possible, it must have already been reached. The aim of the universe can only be eternal life, by drawing its energies from its own resources, eternal generation, the dynamic formations, their destruction and improvement, eternal movement of generation, destruction and eternal improvement, improvement and perfection.

INDIVIDUAL PURPOSE

The eternal forces established in each organism by gravitation and under the influence of the cosmic current of universal life acquire progressively greater knowledge of the world by creating more and more potent mental forces.

UNION WITH THE HIGHER COSMIC CURRENTS: FINAL TRIUMPH OVER GRAVITATION

By the realization of an increasingly closer union with the higher cosmic currents, by more and more perfect adaptation to these currents, attained through utilization of and submission to the laws of nature and to their preconditions of milieu, they progressively refine their material organisms; undergo improvement, and finally *gain freedom from gravitation,* becoming more and more independent of the lowest dynamic formations of matter until their triumph over gravitation is complete.

OPTIMAL COURSE OF ETERNAL AND OMNILATERAL EVOLUTION

They thus maintain their own eternal sources of energy and fall in with the optimal course taken by eternal and omnilateral evolution, passing through increasingly perfect formations.

ETHICS AND RELIGION BASED ON COSMOVITALITY THE AIM OF PANEUBIOTICS

NATURAL AND SIMPLE LIFE – ACTIVE LOVE BASED ON THE VITALITY OF THE COSMIC AND HUMAN RADIATIONS

Individual and Social Law based on Cosmic Law

The aim of paneubiotics is the search for the the omnilateral realization of the best forms of individual, social and cosmic life, in accordance with the following triple laws and optimal purposes:

1. Improvement of self
2. Improvement of others
3. External omnilateral improvement

Individual Purpose: The absorption of every source of energy and harmony from every personal, social and cosmic milieu.

Social Purpose: The molding of the preceding factors into an harmonious balance between the individual and the community, by harmonizing and optimally improving all our individual qualities and values with the realities of our social milieu.

Cosmic Purpose: Harmonization of individual and social balance with omnilateral cosmic law, through the forces and currents of eternal cosmic life.

These three purposes, or objectives, are mutual preconditions and are reciprocally complementary, one of the other.

Social and Moral Psychovitality

Vitality is a power manifested in the universe, spreading to the infinite. It creates, maintains and orders beings who themselves release a limited vitality. The connections between the planets of a solar system show that they are dependent on a common vital center, the sun. The sun, like a nervous ganglion in the human body, is a special seat of vitality, maintaining life in the planets within its province.

Vitality Touches both Living Beings and the Solar System

On the earth, the sun maintains life in plants, animals, and man. But all solar systems themselves are motivated by initial central vitality which, through rotatory movements and force projected from the center of the universe, provides the possibility of life.

Vitality is the Common Property of All Human Beings in Different Degrees; it is derived from the Inexhaustible Source of Life which is Eternal Cosmic Life

Vitality, then, is a radiation which emanates from eternal cosmic life and orders and maintains all the solar systems in space; it is more particularly manifested in the suns, nourishing them continually through the forces of the cosmic radiations which envelop and influence them. This vital force is directed on to the sun and thence to the earth, under different forms, particularly as electromagnetic radiations. At the same time it touches sensible creatures like man, who, being intelligent, should be in perfect harmony with the laws of this vital radiation, which is for him not merely as electromagnetic radiation, but also a source of special benefit in the shape of the human vital radiations. Vitality impinges on the human nervous system in a way which favors the growth and conservation of what is sensible and is consequently drawn into contact with the laws of the vital ocean of the cosmic radiations.

For Vitality to manifest itself continually, Man must observe the Cosmic Law of Perfect Equilibrium

Man, in his turn, can transmit life and nourish things subordinate to him. Vitality in him sustains a perfect equilibrium, which is obtained by the normal functioning of the organism governed by the motor nerves. These nerves are themselves animated by the sensory nerves which cause the whole human machine to experience happiness and well-being. Until the age of science, this human vitality was called "the soul of man." It can maintain its equilibrium in all directions only so long as its existence and forces can be utilized in the normal functioning of the organism within the ambit of the vital radiation and an optimal equilibrium be maintained with the rest of the universe. Such an equilibrium is attained when the senses of a human being are all animated by the cosmic laws. These laws systematically secure to creatures in general and to man in particular what is beneficial to them without any disagreeable consequences being entailed. When this equilibrium is perfect, man is in harmonious contact with the entire universe and with eternal life, from which the effluvia of the cosmic vitality emanate. Therefore man is in himself a microcosmos, a microscopic universe.

Living, in the fullest acceptance of the word, is an art unknown to the human beings of today. They engage in automatic mutual destruction without realizing the nature of their actions, for they are completely lost in the mists of ignorance and egotism.

Real Life, Eternal Cosmic Life, is an Ideal possible to Man

If human beings knew the secret of life and ordered themselves in accordance with the laws of eternal cosmic life, there would be every possibility of their successfully realizing perfect harmony and equilibrium. The human organism is equipped with everything necessary for obtaining through the work of its organs the genial product called Life. Human beings are unaware of their capacity for acquiring eternal life through vitality, though such capacity is completely assured when no obstacle is put either in the way of the functioning of the bodily organs, which operate automatically or in that of the activity of the sensory nervous system linked to the vital radiation which keeps the whole universe in equilibrium and eternal motion. They do not know what true life is, and what it is to live in joy and contentment, where no sorrow, no anxiety is felt, where bitterness, quarrels, disappointment, disease, death are impossible because man as he lives in a sower of good and reaps a rich harvest of happiness.

But as Things are, Man is His Own Enemy

They do not know that every moment of their lives should be happy, because unhappiness is a destructive force in man and robs him of vitality. Happiness, the common goal which men seek, is not reached, because their modes of action, speech and thought are in disharmony with the natural and cosmic law which governs them.

The message this law has for us is a proclamation of active love; it is a secret revealed, a treasure given us by nature; she will give to all who have ears to hear the capacity to become cosmo-vital, to become bio-genic, to receive the impulse of cosmic eternal life through observance of the optimal and omnilateral preconditions on which it depends.

Active Love: the Foundation of all Religions

"Though I speak with the tongues of men and of angels, and have not love, I am become as sounding brass, or a tinkling cymbal.

"And though I have the gift of prophecy, and understand all mysteries, and all knowledge; and though I have all faith, so that I could remove mountains, and have not love, I am nothing.

"And though I bestow all my goods to feed the poor, and though I give my body to be burned, and have not love, it profiteth me nothing.

"Love suffereth long, and is kind; love envieth not; love vaunteth not itself, is not puffed up;

"Doth not behave itself unseemly, seeketh not her own, is not easily provoked, thinketh no evil;

"Rejoiceth not in iniquity, but rejoiceth in the truth;

"Beareth all things, believeth all things, hopeth all things, endureth all things.

"Love never faileth: but whether there be prophecies, they shall fail; whether there shall be tongues, they shall cease; whether there be knowledge, it shall vanish away.

"And now abideth faith, hope, love, these three; but the greatest of these is love."

–I Corinthians, xiii

THE UNIVERSAL FOUNDATION OF ALL RELIGIONS: LOVE

The Perfect Universal Dynamic Unity is Love

Love is manifested as follows:

Love exists *in itself* as the primordial and supreme source of all forms of existence and of all sources of energy, harmony and knowledge. In itself, it is only capable of being lived. It is understandable only in its manifestations, like electricity or heat.

The coexistent dynamic harmony of all *cosmic forces,* factors and laws, makes possible one another's existence and represents one another's sources of energy. Dynamic and eternal harmonious Love in cosmic space is only perceptible to man under the night sky filled with stars and to the astronomer in his observatory. Just as the totality of musical notes is not a symphony without creative genius, so the totality of existing things is not the cosmos without universal creative Love (without which there would be neither correspondence and affinity between Man and the Universe nor, consequently, the possibility of optimal understanding of the omnilateral cosmic laws and forces, nor again, in consequence, the possibility of optimal and omnilateral adaptation).

The coexistent dynamic harmony of all *terrestrial natural* forces and laws (pyrosphere, lithosphere, hydrosphere, botanosphere, atmosphere, nephosphere, stratosphere), makes possible earthly life and serves it as sources of energy, harmony and knowledge (optimal temperature for life, the earth, chemical elements, plants and trees, air, water, sun, movement, changes of season and climate, day and night, the right proportion of all the vital factors, etc.). Love manifested in *terrestrial nature* gives us all that is necessary for health, longevity and happiness. Love is also manifested in nature under several empirical aspects: in physics, as *gravitation;* in chemistry, as *affinity;* in biology, as *vitality;* in psychology, as *consciousness;* in society, as *altruism,* etc.

The coexistent dynamic harmony in all the cells, nerves, organs and senses of the human organism, where each part lives and works for the good of the other parts and of the totality, reveals to us Love manifested in the human organism and the fundamental law of Man: omnilateral, active Love, as the optimal form of human existence in the cadre and under the direction of the totality of the natural and cosmic forces and laws.

Love, manifested in the synthetic and dynamic activity of the *consciousness,* makes it possible for us to know the cosmic, natural, human, social and cultural laws and to employ them as sources of energy, harmony and knowledge for the most perfect realization of our personality and individuality, thereby furthering not only our own individual evolution but also the evolution of the entire planet.

Manifestations:

Love of Life—Optimism

Lack of Love of Life—Pessimism

Love of the Understanding of Life—Wisdom

Lack of such Love—Ignorance, Agnosticism, Skepticism

Love of Action—Activity

Lack of Love of Action—Passivity

Love of Material Forms—Materialism

Love of Ideal Forms—Idealism

Love of Space—Order

Love of Time—Punctuality

Love of the Beautiful—Beauty

In Human Society

The cosmos, nature, the human organism and consciousness, by their primordial law—omnilateral, active Love—also reveal to us the primordial social law: omnilateral, active social Love (altruism) in which individuals represent cells—and society represents the organism. All the parts, therefore, exist for the good of the others and of the totality and, thus, for their own good, because neither in the cosmos, nor in the organism, nor in the consciousness, nor in society, can one part be happy when surrounded by other unhappy parts.

Manifestations:

Exclusive Love of Oneself—Selfishness

Love of Others and Humanity—Altruism

Love of One's Family—Family Love

Love of One's Country—Patriotism

Exclusive Love of One's Country—Chauvinism

Love of One's Class—Class Consciousness

Exclusive Love of One's Class—Class Dogmatism

Love of the Other Sex—Sexual Love

Love of the Old Society—Conservatism

Love of the New Society—Revolutionism

In Human Culture

Human culture, as the higher and omnilater creation of all the preceding factors, has the same laws: omnilateral active Love of all the higher forces united in culture.

Manifestations:

Love of Reality—The Sciences

Love of the Beautiful—The Arts

(in Thought: Literature)
(in Color: Painting)
(in Plastic Forms: Sculpture)
(in Movement: Dance)
(in Sound: Music)

Love of Knowledge—Philosophy

Love of Truth—Morals

Love of Justice—Law

Love of Love—Religion

FUNDAMENTAL COSMOVITAL

Sources, Spheres, Manifestations and Influences of the Rhythmic

			I	II	III
1		Sources of Radiations	Sun		East
2		Radiations	Solar R.	Solar-Terrestrial R.	Terrestrial R.
3		Spheres	Botanosphere	Pyrosphere	Lithosphere
4		Colors	Green	Indigo	Red
5		Sounds	Fa	La	Do
6	TIME	Cycles of the Year	II (Spring)	III (Spring-Summer)	IV (Summer)
7	SPACE	Cycles of the Day	II Morning	III (Forenoon)	IV (Midday)
8		Points of the Compass	Earth	South-East	South
9		Physiological and Psychological Processes	Regeneration Vitality		Construction Accumulation
10		Parts of the Body	Reproductive Systems (Sexual Organs)		Digestive System (Abdomen)
11		Systems of the Organism			Assimilative System (Blood)
12		Cells			Cells of the Red Corpuscles (Haematites)
13		Organs and Senses	Skin (Tactile Sense)		Mouth (Gustatory Sense)
14		Elements			Hydrogen
15		Vitamins	Vitamin "D" and "E'		Vitamin "A"
16		Individual Types	Heliotype	Heliogeotype	Geotype

TABLE OF 128 CORRELATIONS

Cycles of the Cosmic, Solar, Terrestrial and Human Radiations

IV	V	VI	VII	VIII
	Man		Cosmos	
Terrestrial-Human R.	**Human R.**	**Human Cosmic R.**	**Cosmic R.**	**Cosmic-Solar R.**
Hydrosphere	**Anthropo-sphere**	**Atmosphere**	**Cosmosphere**	**Stratosphere**
Orange	**Yellow**	**Violet**	**Blue**	**White**
Re	**Mi**	**Si**	**Sol**	**Silence**
V (Summer Autumn)	**VI (Autumn)**	**VII (Autumn-Winter)**	**VIII (Winter)**	**I (Winter-Spring)**
V Afternoon)	**VI (Dusk)**	**VII (Evening)**	**VIII (Night)**	**I (Down)**
South-West	**West**	**North-West**	**North**	**North-East**
	Clearness Equilibrium		**Superiority Harmony**	
	Respiratory System(Chest)		**Nervous System (Head)**	
	Reactive System(Lymphs)		**Receptive System (Neurons)**	
	White Corpuscles (Leucocytes)		Nerve Cells (Neurons)	
	Nose (Olfactory Sense)		**Eye (Visual Sense)**	**Ear (Auditory Sense)**
	Nitrogen		**Oxygen**	
	Vitamin "C"		Vitamin "B"	
Geoanthro-potype	Anthro-potype	Anthropo-cosmotype	Cosmotype	Cosmo-heliotype

SCHEMATIC AND SYNTHETIC TABLE OF THE UNIVERSE, OF THE EARTH AND OF MAN

Cosmobiotics .... Cosmic (solar) radiations
Geobiotics..Terrestrial radiations
Anthropobiotics..Human radiations

("Cosmovitalism")

A, SPHERES

COSMOS	EARTH	MAN
1. Gaseous celestial bodies (in fiery state).	Pyrosphere (the interior of the earth) (in fiery and gaseous state).	Internal digestive and eliminative assimilative system.
2. Solid celestial bodies (cooled down).	Lithosphere (exterior solid, cooled down shell of the earth).	Solid (solidified) bony system.
3. Spheres of organic life (on certain celestial bodies).	Botanosphere (vegetable kingdom of the earth).	Elastic muscular system.
4. Fluid celestial bodies (in circulation between gaseous and solid state).	Hydrosphere (water of the earth in circulation).	Liquid sanguinous system (red and white corpuscles, etc. in circulation).
5. Cosmic space (circulation of the ether, ultra-ether, etc.)	Atmosphere (air of the earth in circulation, circulation of molecules, atoms, etc.).	Epidermic and pulmonary (surface) system (in the dilation and retraction of the epidermic and pulmonary cells).
6. Cosmic radiation (intermediate con tact in the cosmic environment).	Stratospheric radiations (intermediate contact between the cosmic and earthly environment).	Vibratory system of nervous cells (neurons) (intermediate contact between the organism and the earthly environment).
7. Cosmic ocean of the radiations of the higher cosmic consciences (in the cosmic environment).	Earthly sphere of the radiations of the harmonious and chaotic earthly consciousnesses (in the earthly environment).	Central consciousness of the nervous system (in the human organism).
Dynamic source of "Cosmic Cul ture."	Dynamic source of "Earthly Culture."	Dynamic source of "Individual Culture".

ACTIVITIES

COSMOS	EARTH	MAN
1. Absorption and irradiation of inorganic cosmic energies.	Absorption and irradiation of inorganic earthly energies.	Internal digestion and elimination.
2. Cosmic gravitation of inorganic bodies.	Earthly gravitation of inorganic bodies.	Resistance, solidity, Solidification.
3. Absorption accumulation and irradiation of vital cosmic energies.	Absorption, accumulation and irradiation of vital cosmic earthly energies.	Dynamic concentration of vital human energies.
4. Absorption and condensation of cosmic matter.	Dissolution, assimilation and elimination of earthly matter in the course of its circulation.	Dissolution, assimilation and elimination of bodily matters in the course of their circulation in the organism.
5. Spatial intermediation of cosmic energies.	Spatial intermediation of cosmic earthly energies.	External absorption and intermediation of vital energies and elimination of useless matters.
6. Kinetic propagation of cosmic energies.	Kinetic propagation of cosmic-- earthly energies.	Kinetic propagation of nervous fluxes and energies (reception adjustment and reaction of the neurons).
7. Eternal improvement of the progressive hierarchy of harmonious cosmic processes. Eternal Destruction of chaotic cosmic processes.	Selective struggles between harmonious and chaotic earthly process (under the direction of the cosmic ocean of the higher radiations).	Selective struggle between harmonious and chaotic psychical processes in the central consciousness of the human organism (under the influence of the earthly sphere of radiations).

SIMPLIFIED ANTHROPOCENTRIC COSMOTHERAPY

THE DYNAMIC UNITY OF MAN, THE COSMOS, AND THE THERAPIES

	MAN	COSMOS	THERAPIES
1—	BONES	EARTH	GEOTHERAPY (clay & mud compresses)
2—	BLOOD	WATER	HYDROTHERAPY (Water baths)
3—	SKIN AND LUNGS	AIR	AEROTHERAPHY (air baths-respiration)
4—	NERVES	SUN	HELIOTHERAPY (sun baths)
5—	MUSCLES	VEGETATION	VITAMINO-THERAPY (Diet)
6—	MOVEMENT	MOVEMENT	DYNAMOTHERAPY (exercises)
7—	CELL	ATOM	CELLULOTHERAPY (fasting)

The Twelve Preconditions of an Individual's Degree of Perfection

(from the original *Cosmos, Man and Society*, by the author)

Individual and bodily perfection, health and well-being, necessarily depend on the degree of evolution reached by the twelve known and less-well known senses. This is directly dependent upon the following factors in physical, mental and moral health.

1. Power of accumulating the energy of the eternal cosmic ocean of the higher cosmic radiations.
2. Capacity of power of breathing.
3. Generation of electrical fluid by the expansion of the ganglia of the nervous system.
4. Regularization of the blood circulation in the body by perfect functioning of the lungs.
5. Rhythmic distribution of the blood in the body by the movement of the heart.
6. Equal development of all the cells of the brain.
7. Awakening of the power of thought by the vibrations of the spinal cord.
8. Expansion of the magnetic circuits by the vibrations of the senses of the sympathetic nervous system.
9. Perfect assimilation of food after complete digestion.
10. Purification of the body by its appropriate organs.
11. Generation of vital fluid or regeneration by the organs charged with the conservation of life.
12. Economy in every energy and consciousness of all in existence (optimal and omnilateral orientation).

CREDO

of the International Biogenic Society

We believe that our most precious possession is Life.

We believe we shall mobilize all the forces of Life against the forces of death.

We believe that mutual understanding leads toward mutual cooperation; that mutual cooperation leads toward Peace; and that Peace is the only way of survival for mankind.

We believe that we shall preserve instead of waste our natural resources, which are the heritage of our children.

We believe that we shall avoid the pollution of our air, water, and soil, the basic preconditions of Life.

We believe that we shall preserve the vegetation of our planet: the humble grass which came fifty million years ago, and the majestic trees which came twenty million years ago, to prepare our planet for mankind.

We believe we shall eat only fresh, natural, pure, whole foods, without chemicals and artificial processing.

We believe we shall live a simple, natural, creative life, absorbing all the sources of energy, harmony, and knowledge, in and around us.

We believe that the improvement of life and mankind on our planet must start with individual efforts, as the whole depends on the atoms composing it.

We believe in the Fatherhood of God, the Motherhood of Nature, and the Brotherhood of Man.

—composed in Paris in 1928 by Romain Rolland and Edmond Bordeaux Szekely

APPLICATION FOR ASSOCIATE MEMBERSHIP
INTERNATIONAL BIOGENIC SOCIETY

Please return to: *I.B.S. Internacional*
P.O. Box 849, Nelson, B.C.
Canada V1L 6A5

*Name*____

*Address*______________________________________

*City, State/Prov.,Zip/Code*________________________

*Age*____*Profession*______________________________

*Previous Experience*____________________________

I am interested in:

____*becoming an Associate Member of the I.B.S.*

____*becoming a Teacher of Biogenic Living.*

Enclosed is my annual Associate Membership fee of U.S. $20.00. Please mail my membership card, your current issue of the Periodical Review, *The Essene Way*, and my copy of *The Essene Way-Biogenic Living*, my "Guidebook," textbook, and encyclopedia of ancient wisdom and modern practice.* I understand I will receive a 20% discount on all publications as an Associate Member, but only if I order *directly* from I.B.S. Internacional.

Please make your check in U.S. currency out to
I.B.S. INTERNACIONAL.

P.S. The only reason we ask your age, profession and "previous experience" (whatever that means) is just to get to know you a little, as we may never have the chance to meet you personally. If you don't wish to answer, it's OK.

*If you already have *The Essene Way-Biogenic Living*, please choose any other book from our current catalogue for your I.B.S. membership gift.

RECOMMENDED BOOKS FOR STUDY

Many members who have not yet started study groups or live far from book stores and/or like-minded individuals, are interested in a systematic program of home study. The following books are recommended for such a program, and provide an excellent foundation for study of *The Essene Way of Biogenic Living*, especially when coordinated with the methods outlined in *The Art of Study: the Sorbonne Method.*

PLEASE SEND ME THE FOLLOWING BOOKS:

_____*The Essene Gospel of Peace, Book One $1.00*
_____*The Essene Gospel of Peace, Book Two $9.50*
_____*The Essene Gospel of Peace, Book Three $9.50*
_____*The Essene Gospel of Peace, Book Four $7.50*
_____*The Essene Way-Biogenic Living $11.50 (free with membership)*
_____*The Biogenic Revolution $11.95*
_____*The Chemistry of Youth $9.50*
_____*From Enoch to the Dead Sea Scrolls $7.50*
_____*The Essene Book of Creation $5.95*
_____*The Essene Code of Life $4.50*
_____*The Essene Science of Life $4.50*
_____*Essene Communions with the Infinite $5.95*
_____*Discovery of the Essene Gospel of Peace $7.50*
_____*The Essene Book of Asha $9.50*
_____*Books, Our Eternal Companions $4.50*
_____*Cosmos, Man and Society $8.95*
_____*Search for the Ageless, Volume One $10.50*
_____*Search for the Ageless, Volume Two $11.50*
_____*The Ecological Health Garden, the Book of Survival $7.50*
_____*The Art of Study: the Sorbonne Method $4.50*
________*Sub-Total*
________*Add 15% for Postage & Handling ($3.00 minimum)*

*U.S. $*________________*Total Amount Enclosed*

Please address all orders and correspondence to:

I.B.S. INTERNATIONAL
Box 849, Nelson, B.C., Canada V1L 6A5

*Name*____

*Address*______________________________________

*City, State/Province, Zip/Code*__________________

A MESSAGE FROM SWALLOW

What do you need to know to guarantee a trouble-free journey of Dr. Szekely's "spiritual children," as he called his books, to wherever you live in the world? First of all, in order to receive the special book rate that only the U.S. has, we must have pre-payment in U.S. funds only. Second, if you live in the U.S., just follow the instructions on the last page of our Book List. If you live *outside* the U.S., then BEFORE sending any money, please write to me with a list of the titles you want and any questions you may have. I will answer by return mail, answering your questions, and giving you the exact amount to send, so all you have to do is make your check, money order or bank draft out to IBS INTERNATIONAL. Don't forget: the above address is our *only* address. We send to you and to all those you love, our warmest fraternal greetings of love and peace!